Characterization of Particles in Filter Effluents

Awwa
Research
Foundation

About the Awwa Research Foundation

The Awwa Research Foundation (AwwaRF) is a member-supported, international, nonprofit organization that sponsors research to enable water utilities, public health agencies, and other professionals to provide safe and affordable drinking water to consumers.

The Foundation's mission is to advance the science of water to improve the quality of life. To achieve this mission, the Foundation sponsors studies on all aspects of drinking water, including supply and resources, treatment, monitoring and analysis, distribution, management, and health effects. Funding for research is provided primarily by subscription payments from approximately 1,000 utilities, consulting firms, and manufacturers in North America and abroad. Additional funding comes from collaborative partnerships with other national and international organizations, allowing for resources to be leveraged, expertise to be shared, and broad-based knowledge to be developed and disseminated. Government funding serves as a third source of research dollars.

From its headquarters in Denver, Colorado, the Foundation's staff directs and supports the efforts of more than 800 volunteers who serve on the board of trustees and various committees. These volunteers represent many facets of the water industry, and contribute their expertise to select and monitor research studies that benefit the entire drinking water community.

The results of research are disseminated through a number of channels, including reports, the Web site, conferences, and periodicals.

For subscribers, the Foundation serves as a cooperative program in which water suppliers unite to pool their resources. By applying Foundation research findings, these water suppliers can save substantial costs and stay on the leading edge of drinking water science and technology. Since its inception, AwwaRF has supplied the water community with more than $300 million in applied research.

More information about the Foundation and how to become a subscriber is available on the Web at **www.awwarf.org.**

Characterization of Particles in Filter Effluents

Prepared by:
Jonathan Brandt, **Hélène Lecoanet**, **Ismail Koyuncu**, and **Mark R. Wiesner**
Department of Civil and Environmental Engineering
Rice University, 6100 South Main St., Houston, Texas 77005-1892

and

Srinivas (Vasu) Veerapaneni, **Jessica Edwards-Brandt**, and **Gary Logsdon**
Black & Veatch
8400 Ward Parkway, Kansas City, Missouri 64114

Jointly sponsored by:
Awwa Research Foundation
6666 West Quincy Avenue, Denver, CO 80235-3098

and

U.S. Environmental Protection Agency
Washington D.C.

Published by:

Awwa Research Foundation

American Water Works Association

IWA Publishing

DISCLAIMER

This study was jointly funded by the Awwa Research Foundation (AwwaRF) and the U.S. Environmental Protection Agency (USEPA) under Cooperative Agreement No. R-82940901. AwwaRF and USEPA assume no responsibility for the content of the research study reported in this publication or for the opinions or statements of fact expressed in the report. The mention of trade names for commercial products does not represent or imply the approval or endorsement of either AwwaRF or USEPA. This report is presented solely for informational purposes.

CONTENTS

TABLES

FIGURES

FOREWORD

The Awwa Research Foundation is a nonprofit corporation that is dedicated to the implementation of a research effort to help utilities respond to regulatory requirements and traditional high-priority concerns of the industry. The research agenda is developed through a process of consultation with subscribers and drinking water professionals. Under the umbrella of a Strategic Research Plan, the Research Advisory Council prioritizes the suggested projects based upon current and future needs, applicability, and past work; the recommendations are forwarded to the Board of Trustees for final selection. The foundation also sponsors research projects through the unsolicited proposal process; the Collaborative Research, Research Applications, and Tailored Collaboration programs; and various joint research efforts with organizations such as the U.S. Environmental Protection Agency, the U.S. Bureau of Reclamation, and the Association of California Water Agencies.

This publication is a result of one of these sponsored studies, and it is hoped that its findings will be applied in communities throughout the world. The following report serves not only as a means of communicating the results of the water industry's centralized research program but also as a tool to enlist the further support of the nonmember utilities and individuals.

Projects are managed closely from their inception to the final report by the foundation's staff and large cadre of volunteers who willingly contribute their time and expertise. The foundation serves a planning and management function and awards contracts to other institutions such as water utilities, universities, and engineering films. The funding for this research effort comes primarily from the Subscription Program, through which water utilities subscribe to the research program and make an annual payment proportionate to the volume of water they deliver and consultants and manufacturers subscribe based on their annual billings. The program offers a cost-effective and fair method for funding research in the public interest.

A broad spectrum of water supply issues is addressed by the foundation's research agenda: resources, treatment and operations, distribution and storage, water quality and analysis, toxicology, economics, and management. The ultimate purpose of the coordinated effort is to assist water suppliers to provide the highest possible quality of water economically and reliably. The true benefits are realized when the results are implemented at the utility level. The foundation's trustees are pleased to offer this publication as a contribution toward that end.

Walter J. Bishop
Chair, Board of Trustees
Awwa Research Foundation

Robert C. Renner, P.E.
Executive Director
Awwa Research Foundation

ACKNOWLEDGMENTS

This project was funded by Awwa Research Foundation and USEPA.

The authors of this report are grateful to the following utilities for their cooperation and participating in this project:

Northern Kentucky Water District, Kentucky
Southern Nevada Water Authority, Nevada
City of Cincinnati, Department of Water Works, Ohio
Johnson County Water District No. 1, Kansas
East Water Purification Plant, Houston, Texas
Salem Public Works Department, Oregon
Milwaukee Water Works, Wisconsin
Modesto Irrigation District, California
Buffalo Pound Water Administration Board, Canada.

The authors wish to express their sincere appreciation for the support and encouragement received from Jennifer Warner, project manager for AwwaRF and the following members of the Project Advisory Committee:

Monica Emelko, University of Waterloo
Kim Fox, USEPA
Russell Ford, MWH
Richard Metzger, Monroe County Water Authority.

The authors also thank Gladys Preston of AwwaRF for the assistance in administrative matters.

The authors would also like to thank Heather Landis Mackey, and Nick Burns, Engineers with Black & Veatch for performing sampling at various sites.

EXECUTIVE SUMMARY

INTRODUCTION

It has been nearly 200 years since the first known granular media filtration facility for centralized potable water treatment was put in place in Glasgow, Scotland in 1807. Rapid sand filtration began to be applied here in the United States around 1880; however, widespread application of filtration in drinking water treatment did not occur until the early periods of the 20th century. Today, this widespread Victorian era technology remains as perhaps the most important step in the multi-barrier treatment of drinking water for the vast majority of our nation's population. Despite the widely recognized importance of granular media filtration and an extensive research literature on its operation, we know surprisingly little about the characteristics of particles that enter and subsequently pass through these filters. Of particular concern is the quantity and origin of particles smaller than approximately 0.2 □m, for which conventional particle counting and microscopy are relatively ineffective. Both of these factors are virtually unknown and yet they are critical in diagnosing limitations in filter operation and pretreatment, interpreting the significance of lower turbidity standards, and assessing the impact of these materials on downstream processes. However, particle counting methods currently adapted for online measurement are theoretically blind to these particles. Just as past studies have considered the role of turbidity as an indicator for the removal of larger particles such as bacteria and protozoan cysts, it is prudent to assess the ability of turbidity to act as an indicator of the removal of ultrafine and possibly nanoparticles.

This report provides an in-depth analysis of inorganic and organic constituents in influent and effluent streams to granular media filters at a variety of drinking water treatment plants. The participating utilities are located across the United States and Canada, representing a broad spectrum of pretreatment schemes, filter designs and operation, and raw water sources. The effect of these various parameters on the composition of particles in the filter influent and effluents was evaluated. In addition to characterizing the material composition, the particles were also characterized differentially according to their size distribution across four defined size classes: settleable, filterable, ultrafine, and dissolved +. The dissolved + size fraction includes both truly dissolved materials as well as particles in the nanometer size range. Particular attention was focused on the latter two size classes as they represent materials that are essentially invisible to particle sizing and counting instruments commonly employed at water treatment utilities. The focus on these size classes of particles sets this research effort apart and provides an unprecedented evaluation of materials in granular media influent and effluent streams.

OBJECTIVES

The goal of this project was to develop information that will aid water suppliers in assessing more stringent standards on turbidity, the potential benefits of particle monitoring of filtrate using particle counters, threats posed by ultrafine particles passing through filters, and the impact of pretreatment and filter operation on downstream processes. This study addresses these needs in an unprecedented analytical examination of particles entering and exiting granular media filters that draws on the most recent tools of particle science. The specific objectives of this report are:

1. Characterize the particle composition (inorganic and organic) in filter influent and effluent streams representing a broad range of filtration and pretreatment practices.
2. Differentially characterize particle composition according to size from the micron to nanometer size range.
3. Contrast particle composition as a function of conditions of pretreatment, filter operation, filter type, and source water.

RESEARCH APPROACH

Nine utilities with diverse geographical location, raw water sources, treatment processes, filter designs and operation schemes participated in this study. Most facilities were sampled at two different periods that corresponded to stressed and unstressed operating conditions. Stressed operating conditions were generally defined as periods characterized by high raw water turbidity, organic content and shorter filter runs compared to that during the unstressed periods. Water samples of filter influent and effluent were collected during stressed and unstressed periods for up to three filters. Samples for the filter effluent were collected during the early stages of ripening, the stable period of operation, and immediately before backwashing. Water samples were characterized in terms of total organic carbon, and elemental composition of several inorganic elements. The size distribution of particles in four size classes was analyzed: Settleable ($d \geq 260$ μm), filtrable ($450\text{nm} \leq d < 260$ μm), ultrafine (30 nm $\leq d < 450$ nm) and dissolved + ($d < 30$ nm).

RESULTS AND RELEVANCE TO UTILITIES

Based on the results from this study, a number of conclusions may be drawn regarding the characteristics of particles in influent and effluent streams for granular media filters. Some of the key conclusions are summarized below.

In general, on a mass basis, inorganic and organic materials in the filter influent and effluent were predominantly present in the dissolved + size fraction. Approximately 80% of the inorganic material and 70% of the organic material in the influent and approximately 76% of the inorganic and organic material in the filter effluent are present in the dissolved + size class. The inorganic material in the dissolved + size fraction is primarily composed of hardness cations, and silica and to a lesser extent residual amount of metal coagulants, primarily aluminum. The high fraction of organic material present in the dissolved + size fractions suggest low concentrations, if any, of pathogens in the filter influent as well as effluent for the utilities investigated in this study. While typical size of viruses is within the dissolved + and ultrafine size classes defined in this study, the disinfection process typically achieved high degree of inactivation. The presence of inorganic constituents such as hardness cations, silica and residual amounts of metal coagulants may potentially be of concern for certain downstream processes if required. These may include reverse osmosis process where precipitates of metal silicates could potentially foul the membranes. However, the low concentrations of these ions observed in this study suggest minimal potential for scaling. The relatively high concentration of aluminum observed at some utilities suggests that further optimization of pretreatment process, particularly in response to varying raw water quality is possible.

Materials in the ultrafine size fraction in the filter influent comprised of approximately 14% and 20% of total organic and inorganic material. In the filter effluent, the ultrafine size fraction of these materials ranged from approximately 10 to 20%. Similar to the material present in the dissolved + size range, most of the inorganic material in the ultrafine size fraction is comprised of calcium, magnesium and silica. A strong correlation between the concentration of organic and inorganic materials in the ultrafine size fraction was observed. Even though ultrafine size fraction in filter influent could potentially contribute to headloss development, no direct correlation was observed between concentrations of materials in the ultrafine size fraction and filter run time. However, the data indicates that the potential for shorter filter run times increases beyond loading rates of 200 mg/min/ft^2 inorganic material and 20 mg/min/ft^2 organic material in the ultrafine size class. In most cases, the organic matter concentration in the ultrafine size class increased in the filter effluent compared to the influent suggesting generation of these particles within the filter media and needs further investigation. Ultrafine particle concentrations may be controlled by optimizing pretreatment upstream of filters, and by evaluating filter operating conditions to minimize shearing off of previously formed deposits. While the results indicated that the ultrafine mass fraction is typically low in the filter effluent, it is dependent on the stage of filter run, exhibiting higher concentrations at the beginning of the filter run and end of the filter run.

Traditional turbidity measurements do not reliably indicate the amount of organic and inorganic materials in the submicron size groups (i.e., ultrafine and dissolved + size classes). In general, the effluent turbidity did not correlate well with the amount of material across all size fractions in the filter effluent when considering all of the utilities. The inadequacy of the existing turbidimeters in detecting variations of organic and inorganic material in the submicron size groups limits the degree of optimization that can be practiced in daily operation of facilities. While many utilities have developed several site specific operational procedures to provide safe potable water for their customers, use of more sensitive instruments such as laser turbidimeters can be used to extend the detection limit to submicron particles and should be considered. Ability to detect smaller particles can help optimize operational practices such as filter to waste period and termination of a filter run based on increase concentrations of smaller particles in the effluent.

Residual coagulant metal concentrations were effectively reduced during filtration, although their presence in the filter effluent may contribute to the formation of cakes on the surface of membrane filters that may be placed downstream. No correlation was observed between loading rates of residual coagulants and filter run-time. In general, the results indicated that the coagulation conditions were well optimized at the facilities that were sampled in this study. In some instances, relatively higher concentrations of residual coagulant were observed in the filter effluent that varied during the course of the filter run. Based on the mass removal of various organic and inorganic constituents, it appears that the residual coagulants, despite their lower concentrations, may potentially contribute to significant part of the headloss developed in the filter media.

Results from this study indicated that filter run times tend to increase with increasing Ca/TOC ratios, especially beyond a value of 20. In addition, it was also observed that filter run times tended to decrease with increasing organic loading rates. Further investigation of the effect of organic matter concentrations and its interaction with inorganic material is required to understand its effect on performance of filters.

Practice of pre-chlorinating source water upstream of filters was generally associated with the presence of larger size particles in the filter effluents. The longest filter run-times were observed at facilities that practiced ozonation upstream of filter. Since chlorination also results in the formation of disinfection byproducts, utilities should consider delaying chlorination until after filtration, particularly if it is evident that chlorination is affecting their filter effluent with respect to particle concentrations.

RECOMMENDATIONS FOR FUTURE RESEARCH

Based on the results of this study, further research on the following issues is suggested.

- This study focused on the characterization of particles in the filter influent and effluent. However, to obtain a better understanding of the filter effluent water quality as a function of raw water quality and treatment processes, the water quality through various treatment processes up to the filtration process should also be investigated. In particular, the size distribution of organic and inorganic material through various processes should be investigated. This allows for influencing the size distribution of the organic and inorganic material to optimize filter performance.
- Significant changes to organic and inorganic material size distribution within the filter media were observed a several utilities. For instance, in some utilities, there appears to be generation of ultrafine organic particles within the filter media that resulted in higher concentrations of particles in this size class in the effluent compared to the influent. Further investigation of the balance of organic matter throughout multiple filtration cycles is required to explain these observations.
- The effect of TOC interactions with inorganic material requires further investigation to understand the effect of Ca/TOC ratio effect on filter run time.
- As traditional turbidimeters failed to yield signals that correlated with concentrations of organic and inorganic materials, further investigation is required to identify more advanced instruments such as laser turbidimeters that extend the ability to detect smaller particles.
- Since the bulk of the organic and inorganic material was found to be present in the dissolved + size class, various means of further declassification of this size class should be investigated. In addition, greater resolution of materials smaller than 30 nm, particularly for organic matter, is important to track removal of viruses in the filters. A comparison of removal of viruses of different varieties with inorganic and other organic particles is needed.
- The strong correlation observed in this study between organic and inorganic material in the ultrafine size fraction needs further investigation to identify the nature of interaction, if any, between these two types of particles.
- This study focused on variability between facilities and through time at full scale, which tends to obscure general trends in filter behavior. Investigations at pilot or lab-scale under highly controlled conditions will be required to arrive at a true mechanistic understanding of the interactions between particle size classes within filters, the role of coagulant type on the presence of particles in filter effluents, and the bounds for filter optimization in removing very small particles.

CHAPTER 1
INTRODUCTION

It has been nearly 200 years since the first known granular media filtration facility for centralized potable water treatment was put in place in Glasgow, Scotland in 1807. Rapid sand filtration began to be applied here in the United States around 1880, however, widespread application of filtration in drinking water treatment did not occur until the early parts of the 20th century. The continued development and application of granular media filtration during this time has resulted in a precipitous drop in the incidence of waterborne disease, and has in-turn been the subject of extensive research for the past century (Cutler and Miller 2005; Fox and Lytle 1996; Fitzpatrick 1998; Hatukai, Ben-Tzur, and Rebhum 1997; Tien 1989; Tobiason and Vigneswaran 1994). Cutler and Miller (2005) estimated that with the introduction of filtration and disinfection in several major cities in the US, mortality was reduced by as much as 43% during the period 1900 to 1936. Today, this widespread Victorian era technology remains as perhaps the most important step in the multi-barrier treatment of drinking water for the vast majority of our nation's population. Indeed, a significant number of the reported outbreaks of water borne disease in the United States over the last two decades can be traced to treatment deficiency. For example, during 2001 – 2002, approximately 28% of the recorded 31 outbreaks were attributable to treatment deficiency, while untreated groundwater accounted for 40% of outbreaks, distribution system deficiency contributed to 20% and 12% were due to miscellaneous causes such as contaminated bottled water and storage container (Blackburn et al., 2004). Treatment deficiency included temporary interruption or chronically inadequate disinfection, and inadequate or no filtration. Despite the widely recognized importance of granular media filtration and an extensive research literature on its operation, we know surprisingly little about the characteristics of particles that enter and subsequently pass through these filters. Of particular concern is the quantity and origin of particles smaller than approximately 0.2 □m, for which conventional particle counting and microscopy are relatively ineffective. Both of these factors are virtually unknown and yet they are critical in diagnosing limitations in filter operation and pretreatment, interpreting the significance of lower turbidity standards, and assessing the impact of these materials on downstream processes.

There is abundant practical evidence towards the impact of submicron sized particles on downstream processes, such as membrane systems and disinfection processes (Zhu and Elimelech 1997; Zhu 1996; Zhu and Elimelech 1995; Yiantsios and Karabelas 1998; Schäfer, Fane, and Waite 1998; Nyström, Ruohomaki, and Kaipia 1996; Nyström, Kaipia, and Luque 1995; Li and Elimelech 2004; Hong and Elimelech 1997; Cohen and Probstein 1986). Due to their high diffusivity and ability to form dendritic deposits in porous media, even very low mass concentrations of these materials may over time create high headloss in downstream packed beds such as granular activated carbon (GAC) or ion exchange columns (Tobiason and Vigneswaran 1994, Veerapaneni and Wiesner 1998). Nanoparticles may also form highly compacted and resistive cakes when deposited on membrane filters resulting in higher fouling rates and overall reduced process performance (Cohen and Probstein 1986; Koo, Lee, and Sheikholeslami 2001; Lepore and Ahlert 1988; Nyström, Kaipia, and Luque 1995; Wiesner and Aptel 1996; Zhu and Elimelech 1997). Additionally, as a result of their high surface area to volume ratio, nanoparticles may adsorb disproportionate amounts of contaminants and facilitate their transport through water treatment systems (Masciangioli and Zhang 2003). Additional concerns related to materials in the filter effluents include aesthetic considerations like color and odor issues as well as particles acting as a carbon source for biofilm growth. However, particle counting methods currently adapted for online measurement are theoretically blind to these particles. Just as past studies have considered the role of turbidity as an indicator for the removal of larger particles such as bacteria and protozoan cysts, it is prudent to assess the ability of turbidity to act as an indicator of the removal of ultrafine and possible nanoparticles.

Objectives

The goal of this project is to develop information that will aid water suppliers in assessing more stringent turbidity standards, sources and concentrations of ultrafine particles passing through filters, possible strategies for reducing ultrafine particle concentrations in filter influents and effluents, and the impact of pretreatment and filter operation on downstream processes. This study addresses these needs in an unprecedented analytical examination of particles entering and exiting granular media filters that draws on the most recent tools of particle science.

The specific objectives of this report are:

1. Characterize the particle composition (inorganic and organic) in filter influent and effluent streams representing a broad range of filtration and pretreatment practices.
2. Differentially characterize particle composition according to size from the micron to nanometer size range.
3. Contrast particle composition as a function of conditions of pretreatment, filter operation, filter type, and source water.

CHAPTER 2
BACKGROUND

Amongst the vast technical literature on the topic of particle removal and filtration, there is an important AwwaRF effort that has been published recently. A collaborative effort between AwwaRF and the International Water Supply Association produced the 1998 work, *Treatment Process Selection for Particle Removal*, which provides an overview of particles in source waters, particle measurement techniques, theory of particle removal and unit operation piloting, design, operation and control (AwwaRF-IWSA, 1998). In addition, several AwwaRF research projects addressed various aspects of particle and pathogen removal in the filters (Patania et al. 1996, Huck et al. 2002b, Becker et al. 2004). These efforts focused typically on the relationship between the effluent turbidity, particle removal, and microbial quality of the water. Patania and co-workers (1995) investigated the relationship between particle removal and the removal of protozoan cysts by filters. They concluded that effective removal of cysts was favored when filters were operated to minimize turbidity or particle counts. However, turbidity and particle counts were observed to be more sensitive indicators of filter performance than removal of cysts and they observed very little correlation between the removal of particles greater than 2 μm and the removal of cysts. Subsequent studies (e.g., (Hendricks 2000, Haas et al. 2001) confirmed that proper operation of conventional filters for turbidity removal provides excellent removal of cysts. However, McTigue et al. (1998) found good correlation (R^2=0.7) between particle counts and *Cryptosporidium* oocysts removal, while the removal of *Giardia* cysts was poorly correlated with particles (> 2 μm). Huck et al. (2002a and 2002b) mentioned that the passage of pathogens through filters was found to most likely occur when filters were operated to breakthrough conditions, or when coagulation pretreatment was not optimized to ensure particle destabilization. If the pretreatment efficiency improved, particle removal in filters appeared to improve (Becker et al. 2004).

In summary, the references noted previously indicate that operating filters under conditions that optimize particle removal is one means of maximizing pathogen removal by filtration. However, potential for particle and pathogen breakthrough towards the end of the filter run must also be considered as they may not be detected by increases in turbidity, demonstrating the complexity of the issue. In addition, 'cheap' measurements of particle size and concentration as obtained by devices based on light blockage principle, closely associated time of flight, or particle index methods, although provide a more sensitive alternative to turbidity, are still inadequate for quantifying the amount of materials passing through filters. In particular, these devices are capable of detecting particles larger than 2 μm. While instruments based on Mie theory light scattering principle or an electrical sensing zone (Coulter principle) are capable of measuring smaller particles, they are impractical for the purposes of online monitoring. In addition, it should be noted that instruments based on electrical sensing zone principle can only reliably extend the range of particles measured to no lower than 0.2 μm.

GENERAL CHARACTERISTICS OF FILTER INFLUENT AND EFFLUENT

Materials in filter influents and effluents are either already present in the raw water or added during chemical treatment (Landa, Capella, and Jimenez 1997; Tobiason and Kim 2001; Kim and Hall 2003; Kim and Tobiason 2004). These materials are diverse in origin and include inorganic, organic, and biological materials. Common examples of inorganic materials include iron oxides, silica dioxide, calcites, clays, pure and mixed oxides of aluminum, and many other minerals. Inorganic materials are generally objectionable at sufficient concentrations for aesthetic reasons (e.g., taste, odor, color, and clarity). Inorganic particles also provide substantial surface areas for the adsorption of natural organic matter, soluble organic compounds, and other toxic substances.

In most potable water supplies, the largest fraction of organic material originates from the degradation of plant or animal materials. Organic matter may either be in dissolved or particulate form and can react with oxidants to produce potentially carcinogenic by-products. Organics can be removed through various processes such as physical sieving for larger materials, coagulation/sedimentation processes for smaller materials and advanced processes such as granular activated carbon adsorption, reverse osmosis and nanofiltration (AwwaRF-IWSA, 1998).

Biota commonly found in raw water sources include protozoa, viruses, algae, and bacteria. Pathogenic protozoa such as *Giardia* cysts and *Cryptosporidium* oocysts are of concern because of their prevalence and their higher resistance to chlorine or other oxidants than bacteria and viruses. Algae are unicellular plants ranging in size from approximately 5 to 100 μm. Algae are of concern in potable water treatment for their role in causing taste and odor and their adverse impacts on filter operation and performance. Bacteria are single-celled organisms ranging in size from approximately 0.1 to 5 μm in diameter. Specific bacterial pathogens of concern include *Salmonella, Shigella, Yersinia Enterocolitica, Legionella,* and *Vibrio Cholera*. Additionally, *Escherichia coli* are used as an indicator species for pathogens, thus their presence is also monitored (AwwaRF-IWSA, 1998).

PRETREATMENT

Pretreatment processes such as coagulation, sedimentation, and pre-oxidation are used to maximize the removal of particulate and biological materials prior to filtration (Hargesheimer, McTigue, and Lewis 2000). This has the benefit of improving filter performance and extending filter run-times. Filtration processes are typically classified by the level of pretreatment used, and termed accordingly as conventional, direct, inline, and two-stage filtration. Conventional filtration consists of coagulant addition, flocculation and clarification while direct filtration consists of coagulant addition and flocculation only. The level of pretreatment required is determined by the raw water quality and resources of the operating utility. Inadequate or upsets in the pretreatment processes may result in reduced filter performance, such as shortened run-times and lower removal efficiencies (O'Neill 1964; Roebeck, Dostal, and Woodward 1964; Trussell et al. 1980). Specifically, pH and ionic strength, which are impacted by chemical addition in pretreatment processes, have been identified as significant factors affecting particle detachment in packed bed filters (McDowell-Boyer 1992, Ryan and Gschwend 1994). The type of coagulant used, for example alum or ferric chloride, may also impact filter performance, although not necessarily the removal efficiency of pathogens. For instance, Dugan and Williams (2004) found that *Cryptosporidium* removal was independent of the type of coagulant used. The absence of coagulation processes may also result in reduced pathogen removal (Al-Ani, 1986; Horn et al. 1988; Ongerth and Pecoraro 1995; Patania et al. 1995; Huck et al. 2002a and 2002b).

Pre-oxidation processes have been found to affect the performance of coagulation and filtration processes. These effects vary according to the solution chemistry, oxidant type and entry point, and types of coagulants used. Hatukai, Ben-Tzur, and Rebhum (1997) studied the effect of chemical pretreatment (chlorine and chlorine dioxide) and flocculant dose on the removal efficiency of small (2-5 μm) and large particles (> 5 μm) and found that optimizing their application can delay the observed breakthrough of the larger particles through the filter after ripening. In addition, pilot studies demonstrated that the type of oxidant used and coagulant dosing rate greatly affected particle removal during direct filtration. A recent AwwaRF research study by Becker et al. (2004), studied the effect of oxidants (ozone, chlorine and permanganate) on filter performance and concluded that pre-oxidation can enhance the removal of NOM-coated particles. Oxidation reduces particle stability due to the interactions with and changes to the NOM coating on particles. Chlorine and permanganate can increase the attachment efficiency even in the absence of calcium. Nevertheless, they concluded that the impacts of oxidizing compounds on filtered water turbidity were not substantial. Chlorine has generally been found to be the most effective oxidant, followed by ozone and permanganate, respectively (Huck et al. 2002a and 2002b).

The use of ozone for drinking water treatment in the United States as a pre-oxidant has increased dramatically in recent years due to technological advances and concerns over disinfection by-products. Ozonation has many benefits and its location in the treatment scheme must be chosen by considering its impact on other unit processes, such as filtration (Langlais, Reckhow, and Brink 1991; Becker et al. 2004; Gaubert et al. 2002; Yuksel et al. 2002). For example, better filtered water quality (lower particle counts and turbidity) is observed when ozonation occurs just prior to filtration as opposed to further upstream in the treatment train. A review of the relationship between ozonation, coagulation and filtration processes has been done by Becker and O'Melia (2001). In general, the impacts of ozonation are dependent on the type of coagulant used and the raw water quality. For example, ozone converts NOM into smaller, more oxygenated compounds (e.g., oxalic acid) that may exert a greater metal salt coagulant demand than the parent compounds. Furthermore, because cationic polymers react with particles and large organic matter (and not with the smaller compounds that are formed after ozonation), the optimum polymer dose after ozonation is increased. This may lead to under- or overdosing of coagulant, which can impair the performance of downstream processes. The effect of ozonation increases with increasing hardness to TOC ratios. The ratio of hardness to TOC must be at least 20 before pre-ozonation exhibits significant effect on particle coagulation (Chang and Singer 1988; Langlais, Reckhow, and Brink 1991). To comprehensively assess the performance of granular media filters, with respect to particle removal, run-time etc., it is therefore necessary to consider the impact of pretreatment processes on the filter influent particle characteristics and possibly on the filter itself.

OPERATIONAL FACTORS AND FILTER PERFORMANCE

Particle removal during filtration is a function of the influent particle characteristics, chemical pretreatment, filtration rate, solutions chemistry, and properties of the filter media. Filter performance is affected by both the influent particle size distribution and the change in the size distribution through the filter depth and with time (Mackie and Bai 1992). Particle size distribution is determined by raw water characteristics and pretreatment processes and chemicals used. Particles that are larger than the effective filter pore size are removed through sieving mechanisms. Smaller particles are removed in filters through several mechanisms. The removal of these particles may be considered to be a two-step process – transport to the vicinity of the filter grains and attachment to the filter grains. Depending on the particle size and operating parameters, particles are transported to the vicinity of filter grains through three main mechanisms. Particles with density greater than the fluid deviate from fluid streamlines due to gravity and are transported towards the filter grains. Particles following the streamlines may come in contact with filter grains depending on their size and this mechanism is referred to as interception. Small particles for which diffusive motion is dominant, may be transported to the vicinity of the filter grains by random motion. Once in the vicinity of the filter grains, deposition of the particles depends on the attachment efficiency. Particle attachment to filter media is a function of interfacial chemical interactions between the interacting surfaces, which may be either favorable or unfavorable for attachment. Under unfavorable conditions in which repulsive interactions are predominant, attachment efficiency may be low. Conversely under favorable conditions, the interaction is attractive and particle attachment to the filter media is likely. A key consideration is that as interfacial interaction energies scale with particle size, smaller particles ($d < 100$ nm) will experience comparably different interactions with the filter media compared to larger particles. Hence, filter feed water characterized by large, small or highly polydisperse particle sizes will experience different particle removal efficiencies. In addition, filter run time and headloss development are different for various particle size ranges and concentrations (Lawler 1997). During filter ripening small particles ($d < 1$ μm) in filter influents improve the removal of larger particles, while removal of small particles is not greatly influenced by the presence of larger particles (Veerapaneni and Wiesner 1993). This can be explained by the evolution of deposit morphology as a function of particle size distribution. Smaller particles quickly occupy favorable deposition sites, resulting in highly porous deposits that ultimately reduce the mean filter pore size. This results in increased removal of subsequent particles. This is evidenced by the fact that filter ripening evolves faster when particles in the filter influent have a smaller mass-averaged diameter. Temperature also effects particle removal due to its impact on water viscosity and other properties. Ives and Sholji (1965) found that particle removal decreased with decreasing water temperature with all other variables held constant. The more viscous the water, the more slowly particles move relative to the water to reach a media surface, thereby reducing the removal efficiency.

The filtration rate, defined as the amount of water filtered through a unit cross-sectional area of filter can also influence removal efficiency, by affecting various transport mechanisms, particularly diffusive and gravity dominated mechanisms. High filtration rates also result in higher shear forces causing dislodgment of previous deposited particles. For example, Farizoglu et al. (2003) found that at high filtration rates the rate of headloss over time was lower as a result of reduced particle deposition within the filter. Similarly, Hatukai, Ben-Tzur, and Rebhum (1997) found that filtration rate impacts filter effluent turbidity. It was reported that effluent turbidity decreased with declining filtration rates. They also proposed a relationship between particle removal and retention with filter media size taking into account the higher mass loading rates that result from higher filtration rates. As media size increases, the pore size also increases, resulting in higher shearing forces (Landa, Capella, and Jimenez 1997; Tchobanoglous and Eliassen 1970). For finer filter media, because of the smaller pore sizes, clogging may occur (Tchobanoglous and Eliassen 1970). Large fluctuations in filtration rates have been reported to result in impaired filtered water quality due to particle detachment within the filter (Cleasby, Williamson, and Baumann 1963; Fitzpatrick, Campbell, and Cable 1999). Particle detachment or passage was related to the magnitude and rate of change in the filtration rate and was independent of the duration. It was reported that declining rate filters provide higher particle removals than constant rate filters (DiBernardo and Cleasby 1980, Hudson 1959). However, Hilmoe and Cleasby (1986) reported that constant rate filters performed better than declining rate filters. These contradictory findings suggest that other factors need also be considered.

Particle removal and ultimately the filtered water quality are also dependent on the efficacy of the backwashing procedure employed. Particle detachment from filter media during backwashing is affected by the pretreatment and filtration strategies used (Raveendran and Amirtharajah 1995). For instance, optimization of filtration processes through changes in solution chemistry results in stronger particle attachment to the filter media making it more difficult to remove them during backwashing. If the particles are not fully removed from the filter media during backwashing, they may elute from the filter during start-up and subsequently result in a low quality product (Amirtharajah 1993, Amirtharajah and Wetstein 1980).

Filter media characteristics that affect performance are size (Farizoglu et al. 2003, Trussell et al. 1980), shape, morphology (Trussell et al. 1980, Tufenkji et al. 2004), and surface chemistry (Trussell et al. 1980). The uniformity of the filter media has also been found to influence particle removal. For instance, Yohe et al. (1999) found that the removal of particles in a size range of 3 to 6 μm by anthracite filters improved as the media uniformity coefficients decreased. The size and type of the filter medium determines the total available surface area for particle removal and the size of the flow channels. Initial particle removal will be determined by the number of favorable deposition sites in the filter media and the rate at which these sites become occupied.

MICROBIOLOGICAL REMOVAL

The primary goal of drinking water treatment is to protect public health by providing potable water that is free of microbial and chemical contaminants (Huck et al. 2002a and 2002b). Even with advances in water treatment technologies, outbreaks of water borne disease continue to occur (Fox 2001). The emergence of parasitic protozoa such as *Giardia lamblia* and *Cryptosporidium parvum* as etiological agents of waterborne disease has prompted renewed evaluation of water treatment plant efficiencies. Although disinfection is effective in reducing the threat posed by these parasites, physical removal processes, like filtration, are also important. Optimization of filtration processes for removing these pathogens is further necessary because of growing concerns over disinfection by-products (LeChevalier and Norton 1995, Edzwald et al. 2000). Recently, the Awwa Research Foundation has published several reports on the correlation between pathogen removal and particle counts in filter effluent (Hargesheimer and Lewis 1995; Hargesheimer, Lewis, and Yentsch 1992; Huck et al. 2002b; McTigue et al. 1998; Patania et al. 1995). All of these studies have been used by the drinking water community to assess, install, and maintain particle counters in order to optimize treatment processes for particle and pathogen removal. However, as particle counters are essentially blind to materials in the sub-micron size range further study is required. This is particularly warranted as sub-micron sized particles can inhibit the efficiency of disinfection processes like chlorination and ultraviolet-light irradiation.

There have been several outbreaks of *Cryptosporidium* and *Giardia* which have been linked to inadequate drinking water treatment in the US over the last 20 years (D'Antonio et al. 1985; Hayes et al. 1989; Joseph et al. 1991; Leland et al. 1993; Lisle and Rose 1995; Mackenzie 1994; Moore et al. 1994; Roefer, Monscvitz, and Rexing 1995; Smith et al. 1989). The most notable *Crytosporidiosis* outbreak occurred in Milwaukee in 1993. This particular outbreak was attributed to inadequate chemical conditioning and breakdowns in the granular media filters.

However, outbreaks have been observed at other utilities, such as Las Vegas, Nevada, where there were no apparent treatment deficiencies or breakdowns pointing to the complexity of the problem (Roefer, Monscvitz, and Rexing 1996). These outbreaks have highlighted the need to develop effective treatment barriers for these organisms.

At Milwaukee the filtered water turbidity was measured to be around 2.5 NTU at the time of the outbreak, indicating that a substantial amount of particulate matter was present in the filtered water. The relationship between turbidity and cyst concentration remains incomplete. Nevertheless, because turbidity measurements cannot distinguish the origin or size of the particles present in the water, they could not be directly linked to increased cyst concentrations in the effluent. Several studies have reported a good correlation between reductions in turbidity and particle counts with cyst removal (Fox 2001, Rice et al. 1996, Dugan and Williams 2004) while others have found turbidity and particle counts to be good indicators of filter performance but not cyst removal (Hendricks 2000, Haas 2001, Nieminski and Ongerth 1995). However, McTigue et al. (1998) found good correlation (R^2=0.7) between particle counts and *Cryptosporidium* oocysts removal, while the removal of *Giardia* cysts was poorly correlated with the removal of particles larger than 2 μm. In any event, pathogen removal continues to be gauged by turbidity (Fox and Lytle 1996) and particle counts, despite the imprecise nature of these measurements. This highlights the need to characterize materials that pass through filters and contribute to turbidity, with specific focus on materials in the sub-micron size range. Furthermore, such an analysis would provide a basis from which pathogen removal during filtration may be evaluated.

In filtration, removal of *Cryptosporidium* and *Giardia* is a combination of physical mechanisms (Logsdon et al. 1981, Patania et al. 1995, Tufenkji et al. 2004) and chemically mediated interactions (Hsu, Huang, and Pan 2001; Dai and Hozalski 2002). Removal has been found to be influenced by the solution chemistry (pH and ionic strength) (Hsu, Huang, and Pan 2001) and by the presence of organic materials and biofilm formations (Dai and Hozalski 2002). It is generally concluded that removal decreases with increasing organic concentrations, increasing pH, and decreasing ionic strength. This suggests that charge interactions are significant in cyst removal. Therefore, effective particle destabilization processes may be used to optimize cyst removal during filtration. Additionally, without appropriate chemical pretreatment, filters may not act as an effective barrier against cyst and cyst-sized particles.

CHAPTER 3
EXPERIMENTAL PROCEDURES

PLANT OVERVIEWS

The participating utilities in this study represent an array of raw water sources, treatment schemes, filter designs and operation schemes (Table 3.1). In all, nine utilities took part in this study and were located in the following cities across the United States and Canada: Buffalo Pound, Canada; Cincinnati, Ohio; Lenexa, Kansas; Salem, Oregon; Milwaukee, Wisconsin; Fort Thomas, Kentucky; Modesto, California; Las Vegas, Nevada; and Houston, Texas (Figure 3.1). A number of distinctions may be drawn between the various utilities regarding filter operation (Table 3.1). For instance, the source water for the Milwaukee, Southern Nevada, Buffalo Pound, and Modesto utilities is obtained from a lake or reservoir while the water sources for Johnson County, Salem. Cincinnati, and Northern Kentucky are rivers. The source waters obtained from a reservoir or lake in this study are characterized by low total organic carbon (TOC) and turbidity, while river water sources are characterized by high levels of TOC and turbidity. A second distinction amongst the facilities is the type of pretreatment used prior to filtration. Pre-ozonation is used at the Milwaukee, Modesto and Southern Nevada treatment facilities. At the Milwaukee and Modesto facilities pre-ozonation is employed prior to the rapid mix/flocculation step in the treatment train, while pre-ozonation is applied just prior to direct filtration at the Southern Nevada facility. Pre-oxidation is performed at the Buffalo and Johnson County facilities, using chlorine and chlorine dioxide respectively. In addition, potassium permanganate is used as a pre-oxidant at the Cincinnati, Houston, Johnson County, and Northern Kentucky facilities. The influent pH for the various utilities ranged from 7 to 8 with the exception of Johnson County, a lime softening facility, where influent pH was 9.2. Either aluminum sulfate (alum) or ferric chloride was used as the primary coagulants. Coagulant and filter aids like polyaluminum chloride (PACl) or cationic polymers were also used at some of the utilities. In the following section, a brief summary of general characteristics of each participating utility is given. Additionally, any specific sampling protocols applied to a facility are also noted.

Table 3.1
Characteristics and general operational parameters for the participating water treatment plants

Utility	Source Water	Pretreatment Upstream of Filter	Coagulant	Filter Media Type	Loading Rate (GPM/ft^2)
Milwaukee Water Works, WI	Lake Michigan	Pre-ozonation, coagulation, sedimentation	Alum, Polymer	Anthracite (Depth: 23" Size: 0.855 mm; Uniformity coefficient (UC) <1.32) Sand (Depth: 11" Size: 0.53 mm; UC <1.45)	2.75
Southern Nevada Water Authority, NV	Lake Mead	Pre-ozonation, flocculation	Ferric Chloride	Anthracite (Depth: 19" Size: 1-1.18 mm; UC <1.7) Sand (Depth: 19" Size: 0.42-0.55 mm; UC <1.8) Torpedo Sand (Depth: 1.5" Size: 0.18-0.28 mm; UC <2.2) Gravel (Depth: 12")	6
Johnson County Water District No. 1, KS	Missouri/Kansas Rivers & Groundwater	Rapid mix, pre-sedimentation, oxidation, lime softening, rapid mix, sedimentation	Alum, ferrous sulfate, polymer	Anthracite (Depth: 16" Size: 0.9-1.1 mm; UC <1.5) Sand (Depth: 20" Size: 0.45-0.55 mm; UC <1.6)	5
City of Salem Public water Works, OR	Santiam River	Flocculation	Alum	Sand (Depth 36"; Size: 0.27 – 0.35 mm; UC < 2.5)	0.13 (Slow Sand Filter)

(continued)

Table 3.1 (Continued)

Utility	Source Water	Pretreatment Upstream of Filter	Coagulant	Filter Media Type	Loading Rate (GPM/ft^2)
Buffalo Pound Water Administration Board, Canada	Buffalo Pound Reservoir	Pre-chlorination, flocculation, sedimentation	Alum	Anthracite (Depth: 15"; Size: 1.09 mm; UC < 1.59) Sand (Depth: 15"; Size: 0.515 mm; UC <1.3) Garnet (Depth: 4"; Size: 0.36 mm; UC < 1.43) Gravel support	2.55
City of Cincinnati Department of Water Works, OH	Ohio river	Pre-sedimentation, reservoir storage, oxidation, lime, coagulant addition as needed, secondary sedimentation	Alum, PAC, $Fe_2(SO_4)_3$, Polymer	Sand (Depth: 24-30"; Size: 0.48 mm; UC <1.35)	3
Modesto Irrigation District, CA	Modesto Reservoir	Pre-ozonation, flocculation, sedimentation	Alum, Polymer (winter)	Anthracite (Depth: 72"; Size: 1.5 mm; UC <1.4)	6
City of Houston, TX	Trinity River	Flocculation, sedimentation	Alum	Anthracite/ Sand	6
Northern Kentucky Water District, KY	Ohio River	Pre-oxidation ($KMnO_4$), sedimentation in reservoir, flocculation, and sedimentation	Ferric Chloride, PAC, Polymer	Anthracite (Depth: 16"; Size: 0.95-1.05 mm; UC <1.4) Sand (Depth: 13"; Size: 0.45-0.55 mm; UC <1.6) Garnet (Depth: 4" Size: 0.18-0.28 mm) Gravel (Depth 10")	2.85

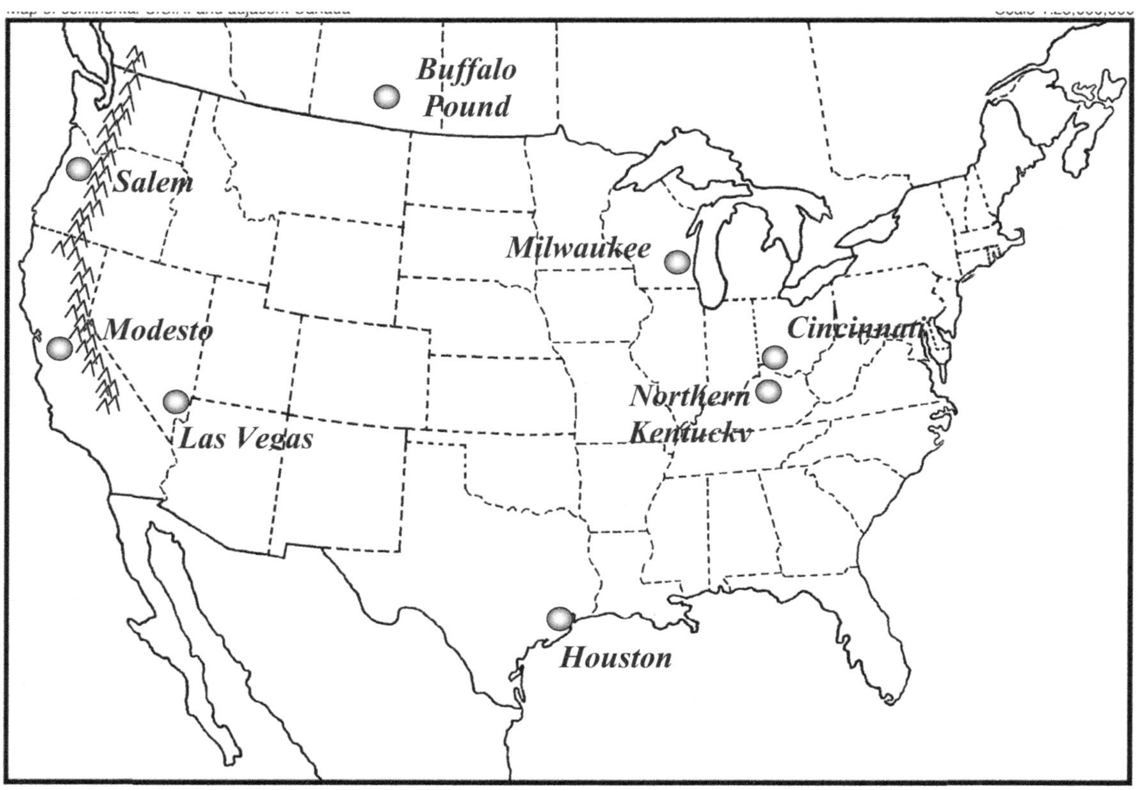

Figure 3.1 Geographical locations of participating utilities

City of Cincinnati, Department of Water Works, Ohio

The City of Cincinnati has three drinking water treatment facilities that include one surface water treatment and two groundwater treatment facilities. The majority of customers (~80%) are serviced by the Richard Miller Water Treatment Plant (RMWTP), the only surface water facility (Figure 3.2). The raw water source for the RMWTP is the Ohio River, whose turbidity ranges between 4 and 1000 NTU, with an average value of around 45 NTU. The RMWTP produces on average 120 mgd of drinking water. The RMWTP treatment scheme consists of pre-sedimentation, coagulation, sedimentation, and filtration processes. In addition, it uses granular activated carbon adsorption downstream of filtration for the removal of organics. The residence time in the pre-sedimentation reservoirs ranges from 2 to 5 days.

Aluminum sulfate ($Al_2(SO_4)_3$) is used as the primary coagulant with an average dose of 15 mg/L. A cationic polymer (1.0 to 3.0 mg/L) is also used to improve solids removal in the sedimentation basins that use lamella plates. Iron sulfate is also used as filtration aid. Lime is used for pH adjustment prior to the filtration step. The influent turbidity to the filters is on average 1.9 NTU. For filtration, the RMWTP utilizes 47 rapid sand filters. Currently, 27 of the filters are composed of gravel, torpedo sand, and 24 to 30 in of filter sand, and a Leopold® clay underdrain. The effluent turbidity from the filters is on average 0.11 NTU. The filters are removed from service and subsequently backwashed when the headloss exceeds 6 ft or the effluent turbidity is greater than 0.3 NTU. If a filter is taken off line for any other reason, the filter is backwashed before being returned to service.

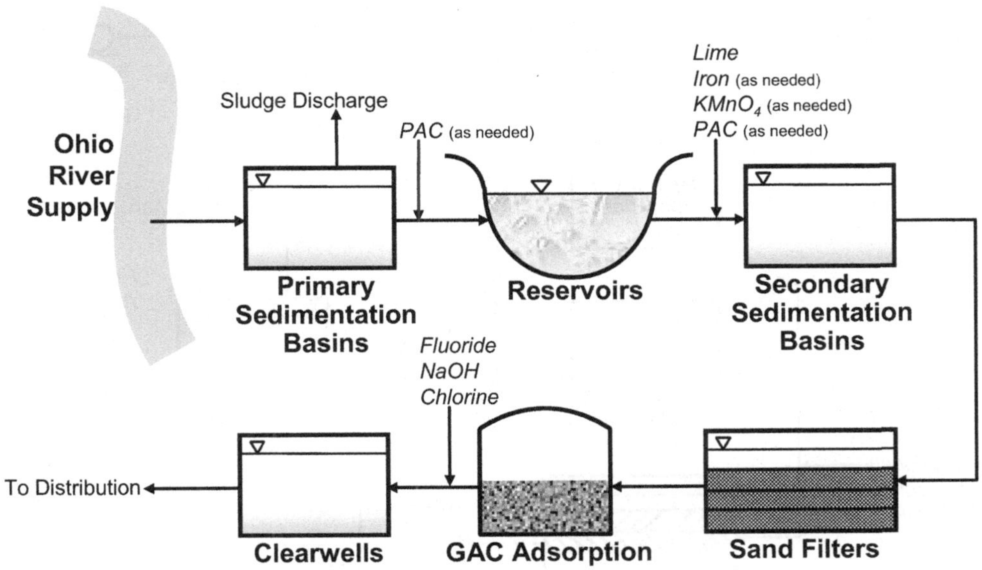

Figure 3.2 Schematic of the Richard Miller Water Treatment Plant, Greater Cincinnati Water Works, OH

Johnson County Water District No.1, Kansas

Water District Number One in Lenexa, Kansas (Figure 3.3) provides drinking water for the greater Johnson County area and has a total treatment capacity of 165 million gallons per day (mgd). Water District One utilizes three water sources - two surface water sources and one groundwater source, which are blended. The surface water sources are the Missouri and Kansas Rivers, while the groundwater source is the Kansas River alluvium aquifer. The Missouri River accounts for approximately 74% of the total raw water for the treatment facility, while the Kansas River and groundwater wells provide 24% and 2% of the total raw water, respectively.

Water District One is a conventional softening water treatment facility. Lime softening is used to reduce the characteristically high hardness of the raw water. Coming out of the first clarification basins, the pH is high due to the lime softening process. Carbon dioxide is thus added in order to lower the pH and make it more stable. Aluminum sulfate ($Al_2(SO_4)_3$) is used as the primary coagulant. Water District One operates 28 dual-media filters, which are composed of anthracite coal and sand. Chlorine dioxide is used as the primary disinfectant, while chloramines act as the residual disinfectant in the storage and the distribution system.

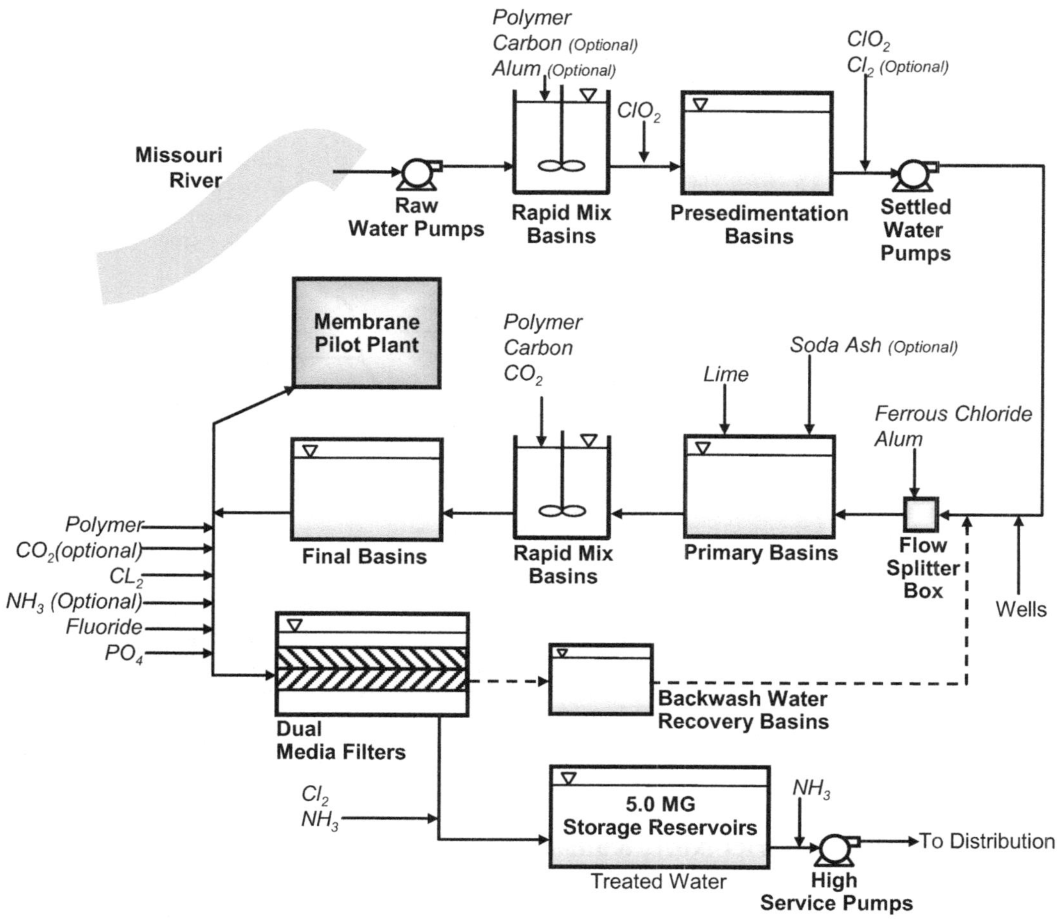

Figure 3.3 Schematic of the Water Treatment Facility 2 of Water District No. 1 of Johnson County, KS

Buffalo Pound Water Administration Board, Canada

The Buffalo Pound utility uses a conventional drinking water treatment scheme incorporating pre-chlorination, alum coagulation, clarification, and filtration (Figure 3.4). Granular activated carbon process is used for about six months during a typical year for taste and odor control. The Buffalo Pound WTP draws its raw water from Buffalo Pound Lake, which is a large shallow lake and is subject to algae blooms. Of particular concern are blue-green algae blooms that occur during the summer months.

Filter media has a depth of about one meter (m) including the garnet sand and gravel. They operate at up to 2.5 m of water above the filter media. Control systems alarm if an individual filter effluent turbidity reaches 0.3 NTU or 2.5 m of headloss and automatically shutdown the filter if effluent turbidity reaches 0.4 NTU. The filter capacity is approximately 18,300 m^3/day or 4.834 mgd. During the sampled filter run, the total filter run time was 45 hours. The total volume filtered over that time period was 17,400 m^3 or 4.623 million US gallons. The plant has an internal filtered water turbidity target of 0.1 NTU or less that is usually met. It is occasionally difficult to achieve in very cold water (< 4 °C).

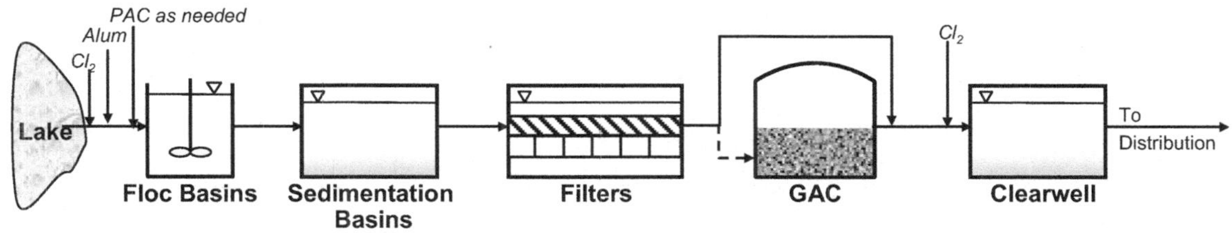

Figure 3.4 Schematic of the Buffalo Pound Water Treatment Plant, Buffalo Pound Lake, Canada

East Water Purification Plant (EWPP), Houston, TX

The East Water Purification Plant (EWPP) is one of four surface water treatment plants, servicing the greater Houston metropolitan area (Figure 3.5). Presently, the EWPP provides approximately two-thirds of the drinking water for all users within the City of Houston. The EWPP utilizes both Lake Houston and the Trinity River as raw water sources. Water from Lake Houston is characterized by high TOC concentrations and turbidity, low alkalinity and low hardness. The quality of water from the lake may change rapidly with weather conditions. Comparatively, water from the Trinity River has lower concentrations of TOC and turbidity, high hardness and high alkalinity.

The EWPP is a conventional drinking water treatment plant. Alum is used as the primary coagulant prior to the sedimentation basins. It utilizes mono-media filters composed of anthracite coal. The filters are 72 in deep with an hour-glass-configured support gravel and Leopold® underdrain. They are operated at 6.0 gpm/ft^2, and have a typical runtime in a range of 36 to 132 hours. The plants are currently operated to achieve a post-filtration turbidity of 0.1 NTU or less

They incorporate air scour for backwashing but do not incorporate filter-to-waste. After backwashing the filters are brought on-line with incremental increases in loading rate to reach operating capacity.

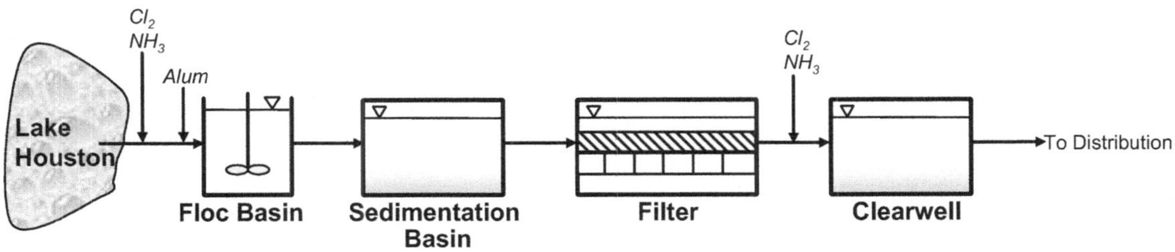

Figure 3.5 Schematic of the East Water Purification Plant, Houston, TX

Salem Public Works Department, OR

Drinking water for the city of Salem, Oregon is provided by the 120 mgd Green Island Treatment facility operated by Salem Public Works Department (Figure 3.6). It serves approximately 177,000 customers. The raw water source for this utility is obtained from the North Santiam River. The raw water is primarily treated using slow sand filtration prior to chlorination and fluoridation. The plant consists of two slow sand filter basins that are subdivided into three filter cells each. The filters are composed of sand supported by a gravel bed. The typical filter run time is between seven to ten weeks.

For this utility only influent and effluent 3-day composite samples were collected. Because the utility uses slow sand filters, no backwash sample was analyzed.

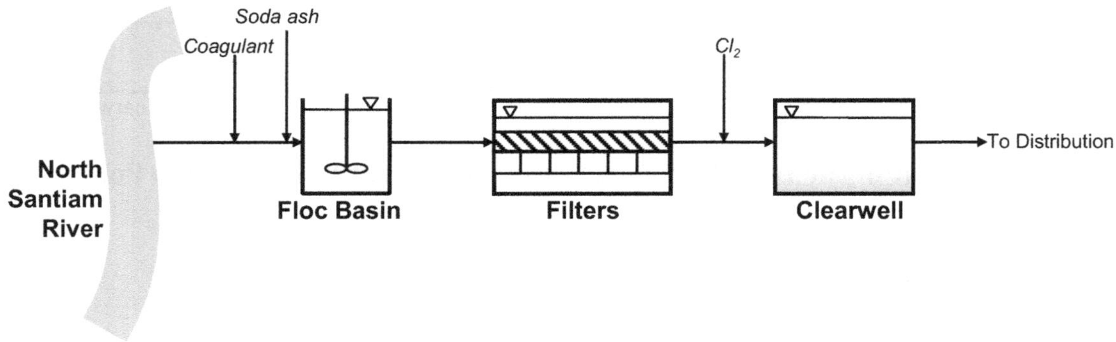

Figure 3.6 Schematic of the Green Island Treatment Facility, Salem, OR

Milwaukee Water Works, Linnwood Plant, WI

The Linnwood Water Treatment Plant is one of two facilities that provide drinking water to the greater Milwaukee area Figure 3.7). Both utilities are operated by the Milwaukee Department of Water Works. The Linnwood plant has a treatment capacity of 275 mgd and draws its raw water from Lake Michigan. The water treatment facility consists of pre-ozonation, coagulation/sedimentation, filtration, and chlorination for disinfection. Alum is used as the primary coagulant along with a polymer as a coagulant aid.

The filters are dual-media and are composed of 24 in of anthracite coal and 12 in of crushed sand with a surface area of 2166 ft^2. The effective size of the anthracite coal is approximately one millimeter (mm) with a uniformity coefficient less than 1.4. The effective size for the sand is between 0.45 to 0.55 mm with a uniformity coefficient less than 1.5. The filters are operated with a constant filtration rate and the plant does not practice filter-to-waste. The average annual run time for the filters is between six to eight days.

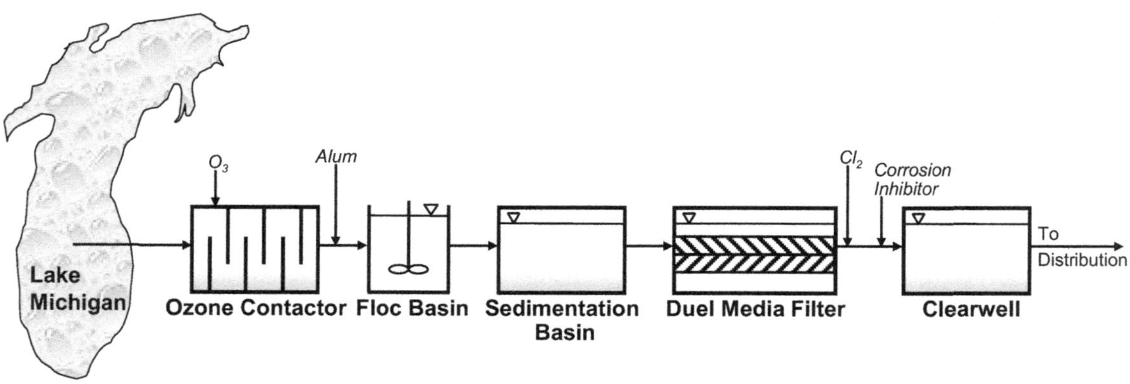

Figure 3.7 Schematic of the Linnwood Plant, Milwaukee Water Works, WI

Northern Kentucky Water District, Fort Thomas Plant, KY

The Northern Kentucky Water District operates a 44 mgd conventional water treatment plant in Fort Thomas, Kentucky that treats water from the Ohio River (Figure 3.8). A blend of alum and polymer is used as the primary coagulant and ferric sulfate is used as coagulant aid.

The utility operates twelve dual media filters composed of anthracite coal (16 inches deep, 1 mm effective size and uniformity coefficient of 1.40) and sand (13 inches deep, 0.5 mm effective size and uniformity coefficient of 1.60), each with a cross-sectional area of 560 ft^2. The underdrain is composed of garnet (4 inches deep, effective size 0.23 mm) and gravel (10 inches deep). The filters operate at an average loading rate of 2.85 gal/ft^2/min, with typical run time duration between 60 to 100 hours. The filters are backwashed at 100 hours typically, or when the terminal headloss or effluent turbidity exceeds set limits.

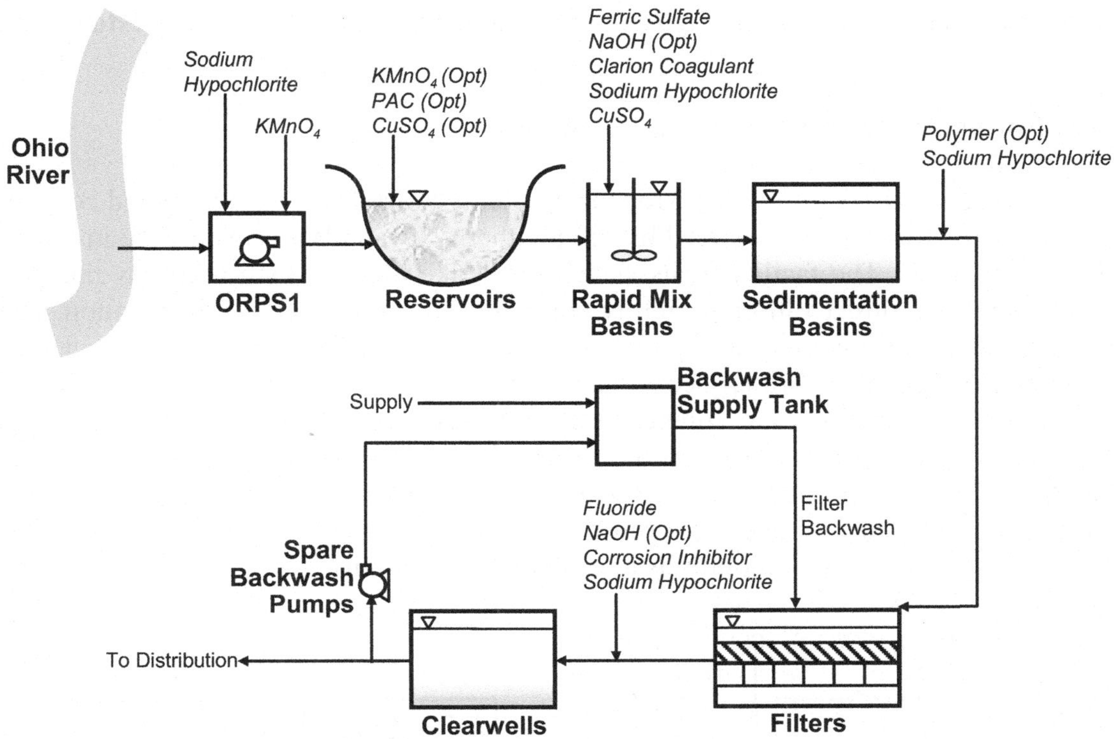

Figure 3.8 Schematic of the Fort Thomas Water Treatment Plant, Northern Kentucky Water District, KY

Modesto Irrigation District, CA

The Modesto Regional Water Treatment Plant (MRWTP) is a 30 mgd facility that can be operated in either a conventional or direct filtration mode (Figure 3.9). The primary raw water source for the facility is the Modesto Reservoir. The MRWTP incorporates pre-ozonation, dual-stage coagulation, tapered-velocity gradient flocculation, sedimentation, pre-filtration hypo-chlorination, filtration, post-filter hypo-chlorination, and stabilization utilizing hydrated lime, carbon dioxide and sodium hydroxide.

The filters are 72 in deep-bed mono-media anthracite coal (6 ft) filters with hour-glass-configured support gravel (14 in) and Leopold® under drain. They have a cross-sectional area of 204 ft^2 and are typically operated at 6.0 gpm/ft^2. The media size is 1.5 mm effective size with a uniformity coefficient of <1.4. The typical filter run time is between 36 to 132 hours. Filters are typically backwashed after 144 hours of operation or if the headloss is eight feet or higher. A backwash may also be initiated when particle counts are high for a given filter and they continue to rise, indicating possible breakthrough. The filters are backwashed using air scouring and there is no filter-to-waste practiced.

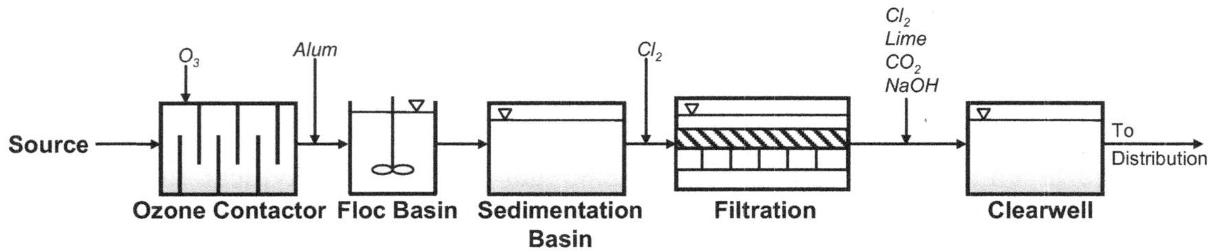

Figure 3.9 Schematic of the Modesto Regional Water Treatment Plant, Modesto, CA

Southern Nevada Water Authority, Las Vegas, NV

The Alfred Merit Smith Water Treatment Facility (AMSWTF) is a 600 mgd conventional drinking water treatment plant operated by the Southern Nevada Water Authority in Las Vegas, Nevada (Figure 3.10). The facility treats water from Lake Mead, which is fed by the Colorado River. The plant is a direct filtration facility consisting of pre-ozonation, flocculation, filtration and post-chlorination. This plant utilizes pre-ozonation for disinfection prior to direct filtration. Ferric chloride is used as the primary coagulant.

The dual-media filters are composed of anthracite coal (19 in) and sand (9 in), with a support layer of garnet (2 in). The filters are operated at an average rate of 5 gpm/ft^2; however, a variable rate (2.5 to 5 gpm/ft^2) is typically used throughout a given day to coincide with the production required. The average filter run time is approximately 70 hours. Shorter filter run-times occur during the summer months as a result of algae blooms.

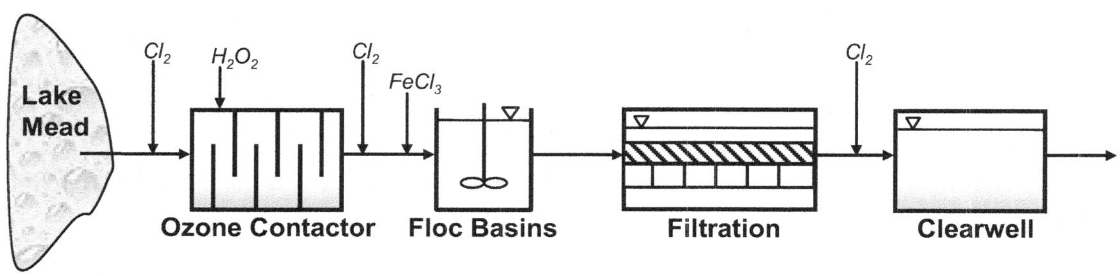

Figure 3.10 Schematic of the Alfred Merritt Smith Water Treatment Facility, Boulder City, NV

EXPERIMENTAL METHODS

Detailed descriptions of the sampling methodologies and experimental protocol used in this investigation are included in Appendix A and graphical sampling, processing, and labeling procedures are also presented there.

Size Fractionation

Particles in the water samples were fractionated according to size using the methodology developed by Grout, Wiesner, and Bottero (1999). The organic and inorganic materials were fractionated into four different size groups, defined as follows:

Settleable: d ≥ 260 μm
Filtrable: 260 μm > d ≥ 450 nm
Ultrafine: 450 nm > d ≥ 30 nm, and
Dissolved +: d < 30 nm.

The settleable fraction was separated by allowing a sample to settle for 30 min and withdrawing the supernatant. The remaining solution was then assumed to contain particles both larger and smaller than 260 μm while the supernatant contained only particles smaller than 260 μm (i.e., filterable, ultrafine and dissolved + fractions).

The combined ultrafine and dissolved + fractions were then separated from the filtrable fraction by passing the supernatant through a 0.45 μm cellulose nitrate filter (Sartorius AG, Goettingen, Germany). Prior to use, each membrane was washed with a 10% HCl solution, rinsed with ultrapure water and then dried at 37°C.

The dissolved + fraction was then separated from the ultrafine fraction using an ultracentrifuge (Beckman Coulter Optima LE-80K, Fullerton, CA). Ultracentrifugation of the filtrate was carried out at 30000 rpm for 16 hours and at 10°C. Using Stokes' law the particle size in the supernatant after ultracentrifugation is calculated to be below 30 nm.

A summary of the fractionation procedure used is provided in Figure 3.11. From the size fractionation procedure used in this study, a relative size distribution based on the concentration of materials in the different groups is developed. However, the particle size distribution discussed here is not the same as that measured using for example dynamic light scattering. The distribution as discussed here describes the amount of materials in size groups and does not provide size distribution within each group.

Sampling and Analysis

The facilities were sampled at two different periods that corresponded to stressed and unstressed operating conditions. Stressed operating conditions were generally defined as periods characterized by high raw water turbidity, organic content, and shorter filter runs compared to that during the unstressed periods. These two periods were identified for each utility based on historical data for the respective facilities. The Buffalo Pound and Salem facilities were however, only sampled during the stressed operating period. Water samples were collected from the filter influent, effluent, and during the backwash event for up to three different filters. Samples from the filter effluent were collected during the early stages of ripening, the stable period of operation, and immediately before backwashing. This allowed evaluation of the filter effluent quality at various stages of a filter, namely, ripening period, steady state operation and end of filter run.

Water samples were characterized in terms of total organic carbon (TOC) and elemental composition of several (major and trace elements. Total organic carbon was measured using a Shimadzu TOC analyzer model 5050A (Columbia, MD). For TOC analysis the water samples were dosed with sodium azide (1 mg/L) to inhibit biological activity and were stored in brown glass bottles. Elemental characterization was carried out using an inductively coupled plasma optical emission spectrometer (ICP/OES, Perkin Elmer Instrument Optima 4300DV, Wellesley, MA) and with an ICP mass spectrometer (ICP/MS, Perkin Elmer Sciex Elan 9000) for trace analysis. All chemical data were calculated by subtracting a blank-reference difference from the measured value. The elements that were analyzed for included silica (Si), calcium (Ca), aluminum (Al), iron (Fe), manganese (Mn), magnesium (Mg), copper (Cu), zinc (Zn), lead (Pb), and arsenic (As). Samples for elemental analysis were stabilized with nitric acid at pH below 2 and were stored in Teflon bottles.

Morphological Characterization

Particles in the filtrable and ultrafine size fractions were further characterized using scanning electron microscopy (SEM) and transmission electron microscopy (TEM) to determine their structural characteristics. Filtrable materials were analyzed through SEM imaging of the deposit retained on the 0.45 μm filters using an environmental scanning electron microscope (JEOL JSM 5300 ESEM, Peabody, MA). Elemental chemical analysis of the filter cake was done using an energy dispersive spectrometer (EDS) and analysis system X-rays (EDXRA). These measurements provided a relative description of the elemental composition of the filtrable fraction in addition to showing the structural characteristics of some of these materials. For the TEM analysis the ultrafine particles were first deposited on Formvar Carbon Copper 200-mesh microgrids (Structure Probe, Inc., West Chester, PA) by horizontal ultracentrifugation separation for 16 hours at 25000 rpm and 10°C. The particles were then imaged using a transmission electron microscope (JEOL 2010 FasTEM, Peabody, MA) operating at 100 kV.

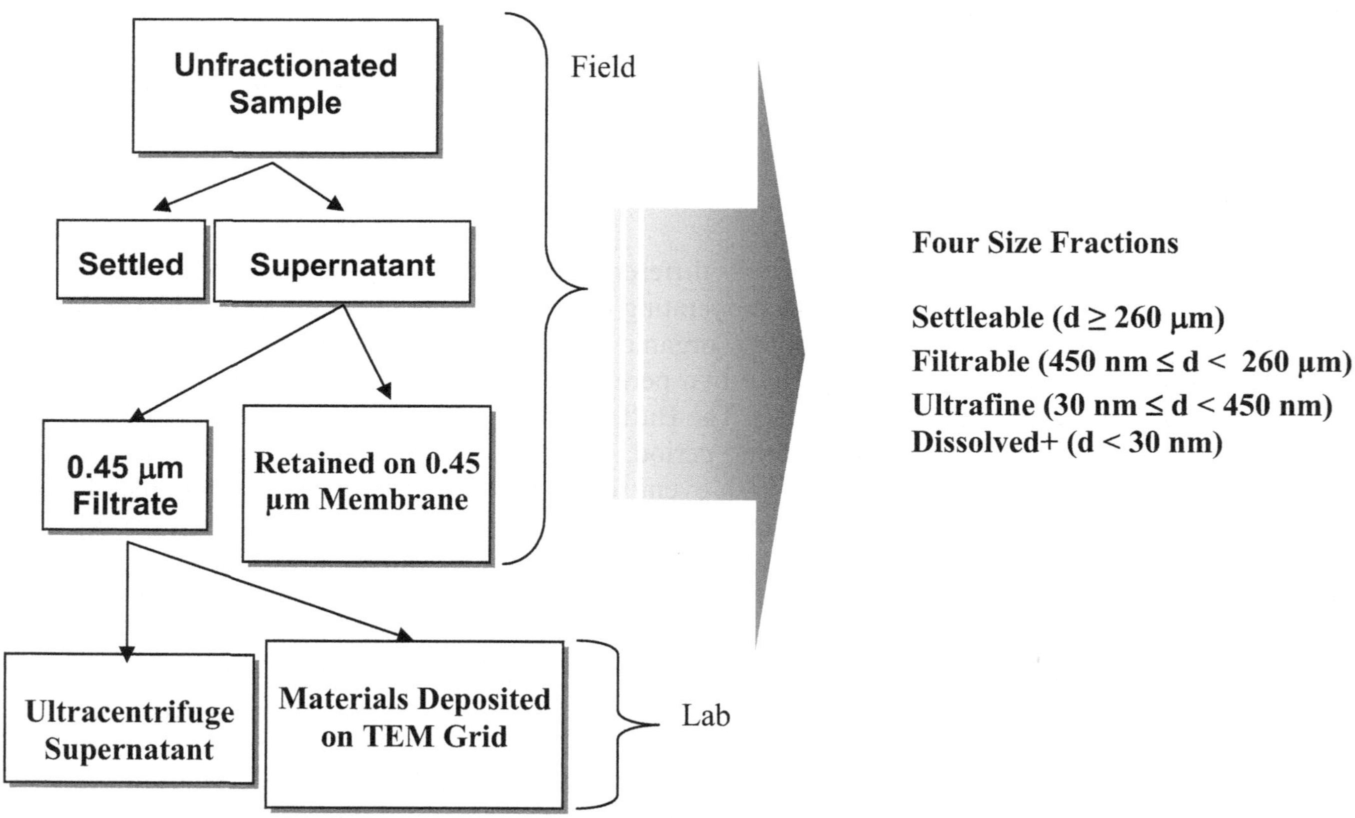

Figure 3.11 Sequence for fractionating particle suspensions according to size

Drinking Water Regulations

Drinking water standards, as established by the USEPA, which are particularly relevant to the current investigation, are summarized in Table 3.2.

Table 3.2
Drinking water standards relevant to the current investigation as established by the USEPA

	Constituent	MCL or TT or MRDL (mg/L unless stated otherwise)
Primary Standards	Arsenic	0.010
	Barium	2
	Copper	TT – Action level 1.3
	Lead	TT – Action level 0.015
	Turbidity	100% of samples < 1 NTU 95% of samples < 0.3 NTU
Secondary Standards		
	Aluminum	0.05-0.2
	Chloride	250
	Color	15 color units
	Copper	1.0
	Fluoride	2.0
	Iron	0.3
	Manganese	0.05
	pH	6.5-8.5 standard units
	Total Dissolved Solids	500

Reference: http://www.epa.gov/safewater/mcl.html
Notes:
MCL – Maximum contaminant level
TT – Treatment technique - a required process intended to reduce the level of a contaminant in drinking water
MRDL – Maximum residual disinfectant level
Action Level – A treatment technique required if more than 10% of tap water samples exceed the action level
Secondary standards – Non-enforceable guidelines regulating contaminants that may cause cosmetic effects or aesthetic effects. States may choose to adopt them as enforceable standards.

CHAPTER 4
RESULTS

The results of filter influent and effluent analyses for all the utilities investigated in this study are presented in this chapter. Further discussion of the results is continued in Chapter 5.

MILWAUKEE WATER WORKS - LINNWOOD WATER TREATMENT PLANT

Mass size distribution of organic and inorganic material present in the Linnwood water treatment plant's filter influent and effluent during stressed and unstressed periods is given in Figure 4.1 – 4.4. The organic mass size fractions for the filter influent and effluent are shown in Figure 4.1 during the stressed filter period. For the influent, almost all of the organic material (> 99%) was in the dissolved + size fraction. In the filter effluent, the organic material is distributed approximately in the range of 50% dissolved +, 32% ultrafine, 2% filterable and 14% settleable fractions. It is interesting to note that the organic material distribution between these size classes remained mostly unchanged during various stages of filter run (Figure 4.1b, 4.1c and 4.1d). The increase in the settleable, filtrable, and ultrafine organic size fractions in the effluent relative to those measured in the filter influent suggests that aggregates are being released by the filter. This may be expected after filter maturation, but not at the beginning of the filter run when the filter is clean, as observed here. The reason for this is not clear.

The distribution of metals into various size fractions for the filter influent and effluent under stressed filter conditions are shown in Figure 4.2. Metals in the filter influent (Figure 4.2) are largely concentrated in the dissolved + size fraction. Similarly, the filter effluent is characterized almost completely by metals in the dissolved + size fraction at each filter stage. Elemental analysis of the particles revealed that the bulk of the metals are calcium and magnesium, with lower concentrations of silica and aluminum. Similarly, calcium and magnesium compose the bulk of particles in the dissolved + size fraction with lower amounts of silica.

The distribution of metals into relative size fractions for the filter influent under unstressed filter conditions is reported in Figure 4.3. Metals in the filter influent are principally present in the dissolved + size fraction followed by the settleable, ultrafine, and filtrable fractions, respectively. Compared to the stressed period, higher percentage of metals are present in the settleable fraction during unstressed period. Calcium and magnesium are the principle metallic species in the filter influent followed by aluminum, silica, and iron, respectively. Similarly, calcium and magnesium compose the bulk of particles in the dissolved + size fraction again followed by aluminum, silica, and iron particles.

The organic particle size fractions for the filter influent and effluent are reported in Figure 4.4 for the unstressed filter conditions. From Figure 4.4a, organics in the filter influent are present primarily in the dissolved + and settleable size fractions, with lower concentrations in the filterable, ultrafine, and dissolved + fractions. The effluent is similarly characterized at the start, middle, and end of the filter run (Figure 4.4b, c and d) where the organic material is largely concentrated in the dissolved + size fraction (~95%).

SEM images of influent and effluent retained on a 0.45-µm filter pad at the Milwaukee Water Works water treatment facility are shown in Figure 4.5. SEM images of the influent deposit show filter clogging (Figure 4.5a). From SEM/EDX analysis, composition of fouling layer is 94% Al (Al_2O_3), 4% Ca (CaO), and 2% Fe (Fe_2O_3). No deposit was observed from effluent water filtration and the surface was similar to a clean filter-pad (Figure 4.5b). In addition, TEM image of the filter effluent is given in Figure 4.6.

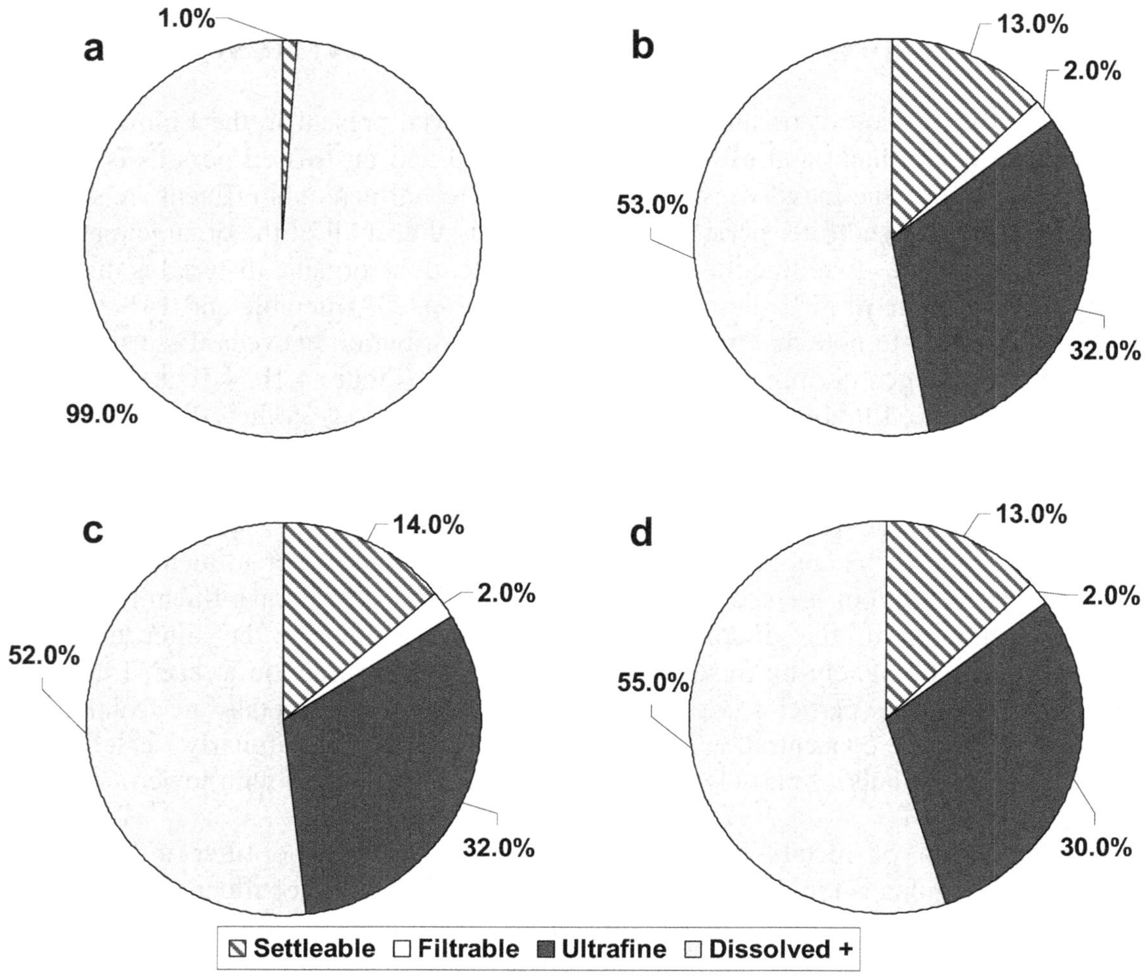

Figure 4.1 Size fractionation of the filter (a) influent and effluent TOC at the (b) start, (c) middle, and (d) end of the filter run for the Milwaukee Water Works water treatment facility during the stressed sampling period (influent $[TOC]_T$ = 2.39 ppm, effluent $[TOC]_T$ = (a) 3.15 ppm, (b) 3.21 ppm, and (c) 3.25 ppm)

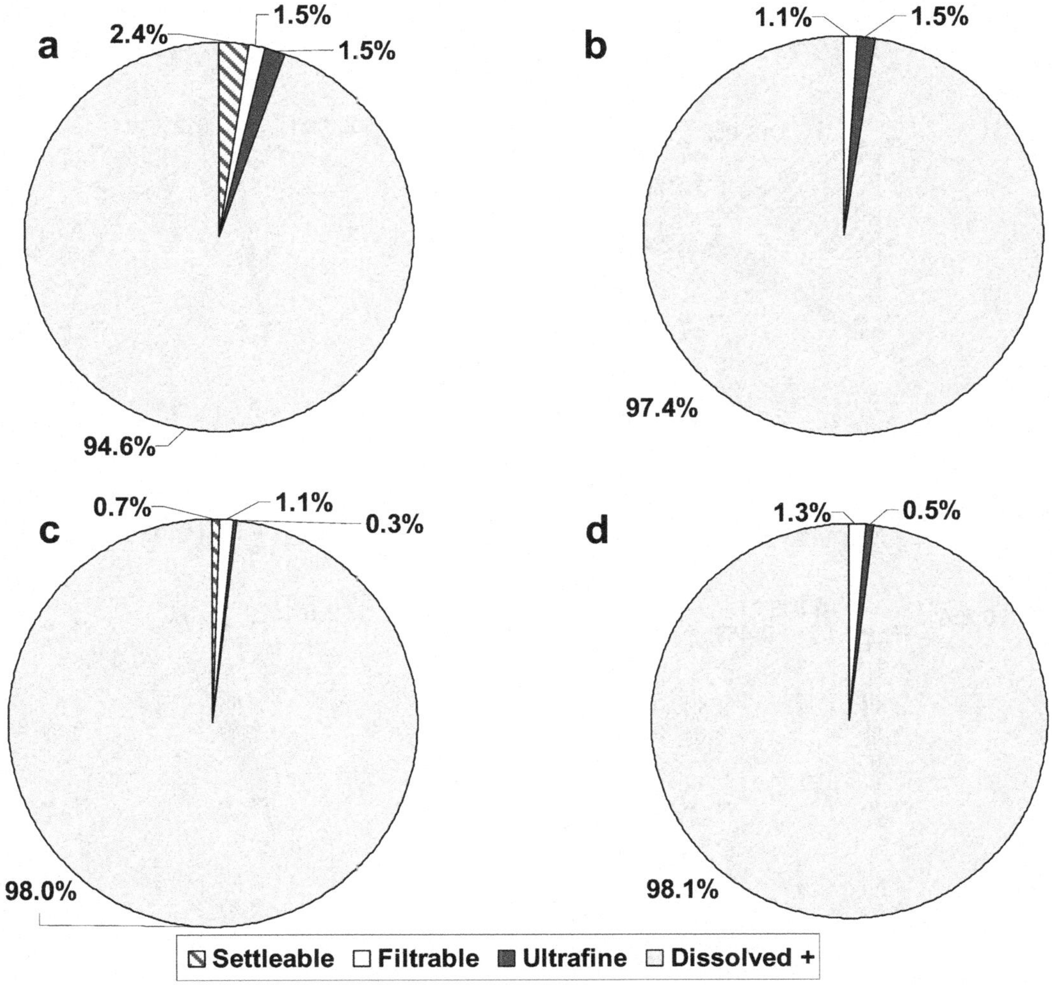

Figure 4.2 Size fractionation of the filter (a) influent and effluent elemental composition at the (b) start, (c) middle, and (d) end of the filter run for the Milwaukee Water Works water treatment facility during the stressed sampling period (influent $[Metals]_T$ = 50.52 ppm, effluent $[Metals]_T$ = ((b) 48.98 ppm, (c) 48.59 ppm, and (d) 49.20)

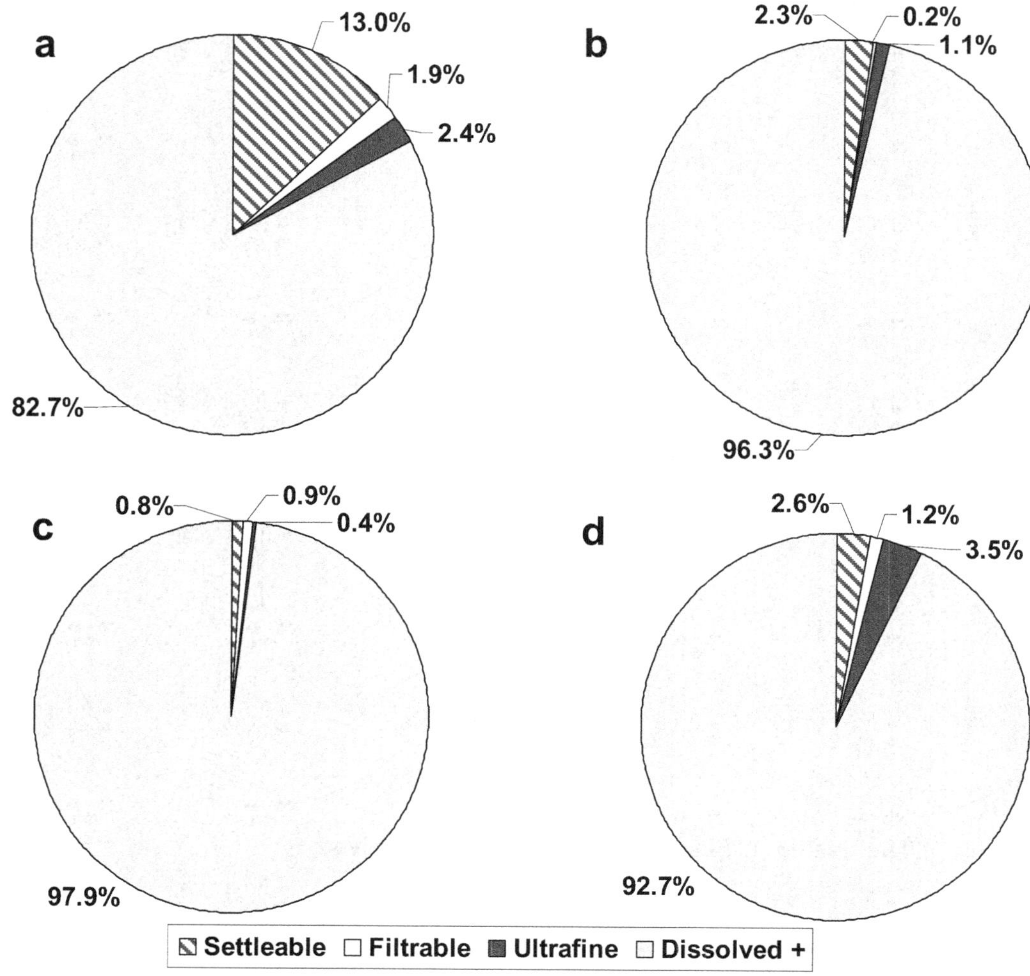

Figure 4.3 Size fractionation of the filter (a) influent and effluent elemental composition at the (b) start, (c) middle, and (d) end of the filter run for the Milwaukee Water Works water treatment facility during the unstressed sampling period (influent $[\text{Metals}]_T$ = 45.95 ppm, effluent $[\text{Metals}]_T$ = 48.62 ppm)

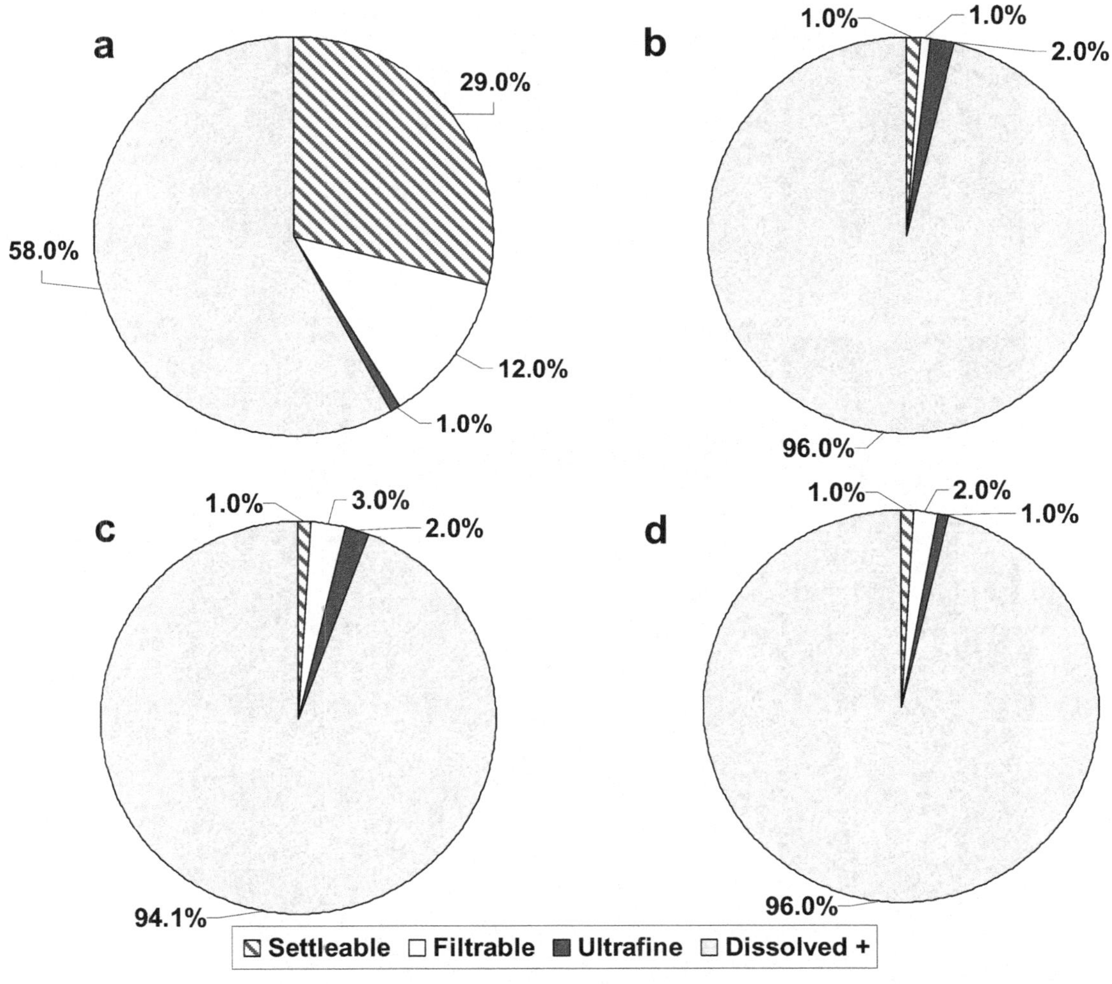

Figure 4.4 Size fractionation of the filter (a) influent and effluent TOC at the (b) start, (c) middle, and (d) end of the filter run for the Milwaukee Water Works water treatment facility during the unstressed sampling period (influent $[TOC]_T$ = 1.82 ppm, effluent $[TOC]_T$ =0.82 ppm)

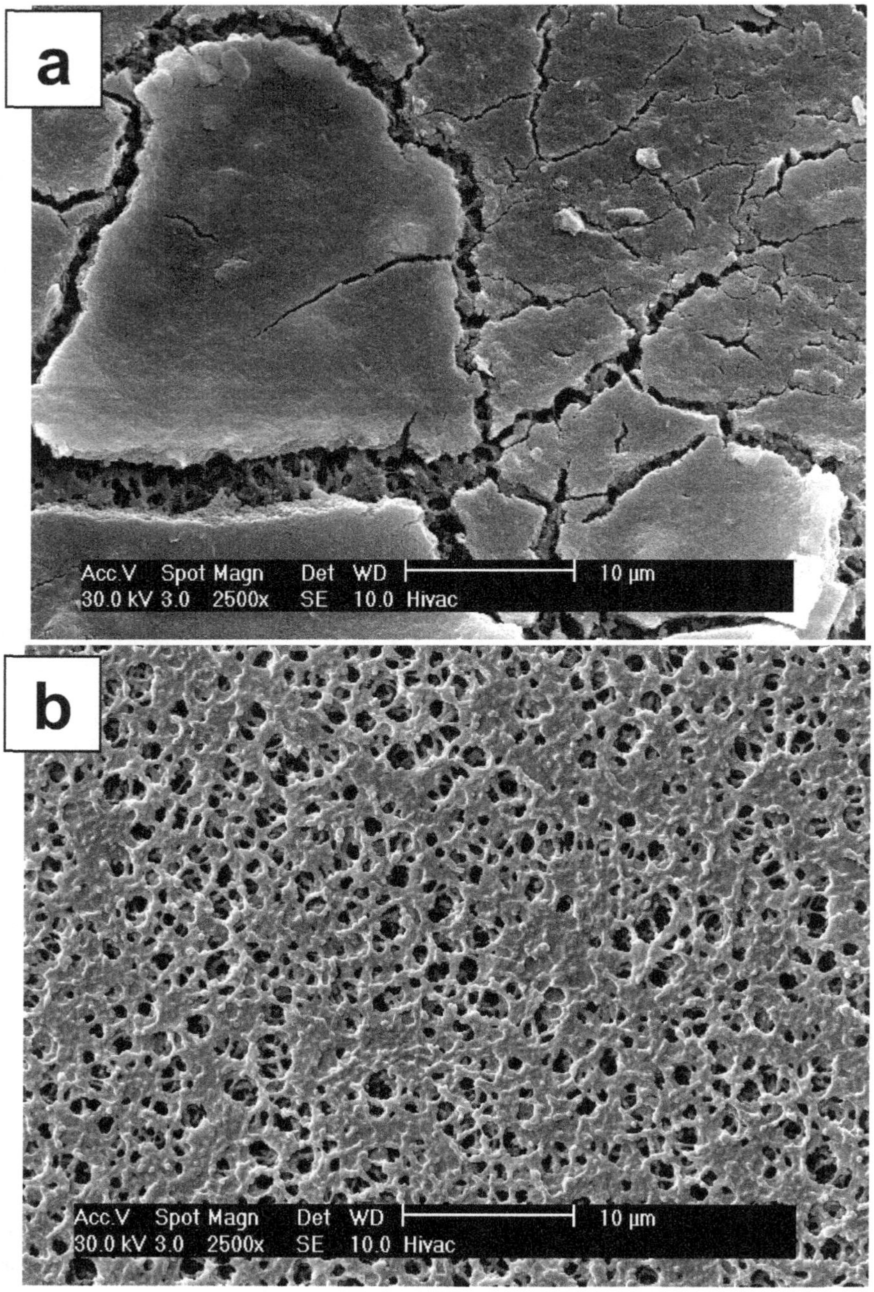

Figure 4.5 SEM images of material from the filter (a) influent and (b) effluent retained on a 0.45-µm filter pad at the Milwaukee Water Works water treatment facility during unstressed period. The effluent sample was collected from Filter #13 during the middle of the filter run. The composition the influent cake is 94% Al (Al_2O_3), 4% Ca (CaO), and 2% Fe (Fe_2O_3) based on EDAX analysis

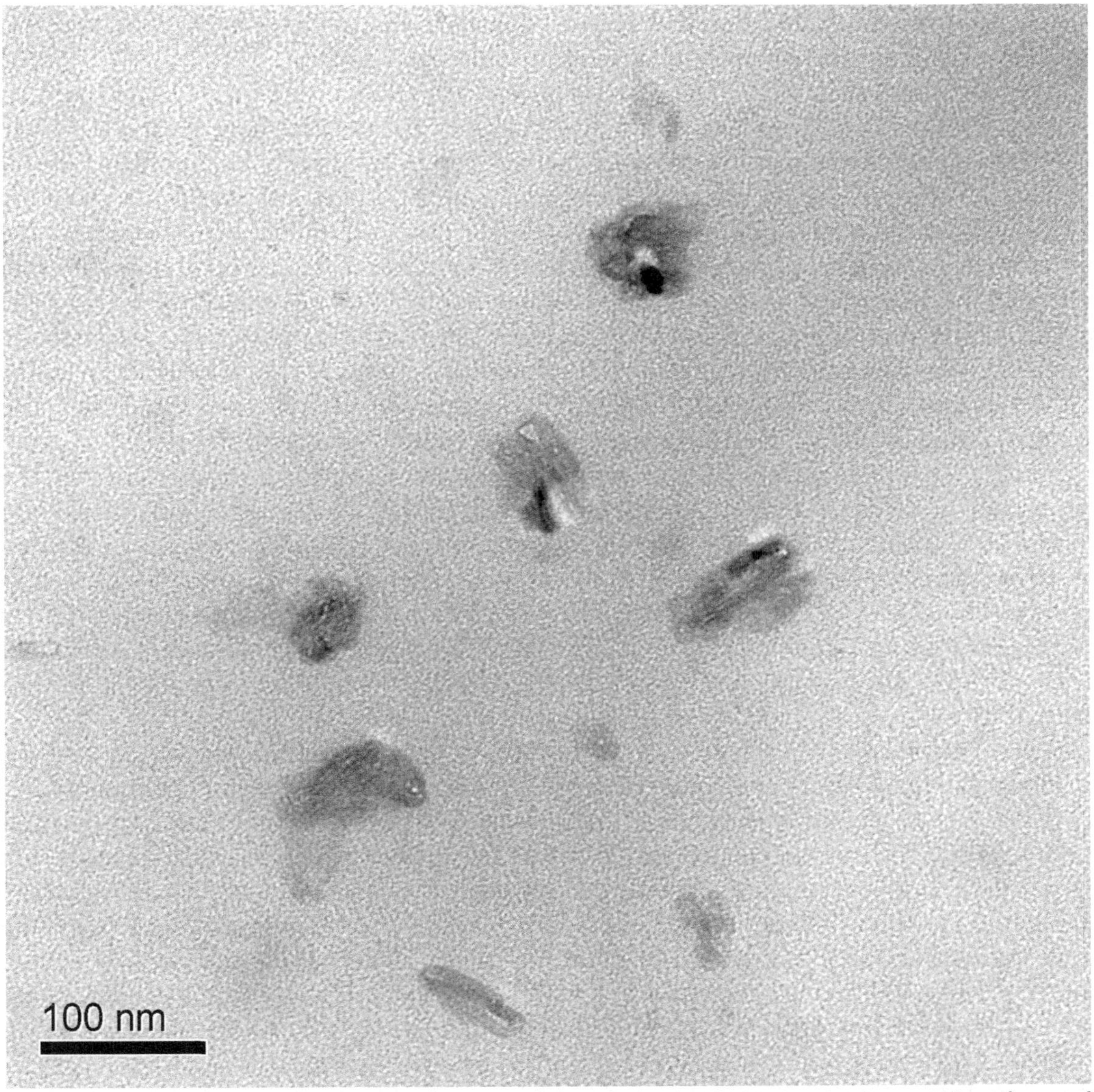

Figure 4.6 TEM image of ultrafine particles deposited on a copper mesh microgrid from the filter effluent at the Milwaukee Water Works water treatment facility during unstressed period. The effluent sample was collected at Filter #16 at the start of the filter run

Northern Kentucky Water District, KY

Size distributions of organic and inorganic constituents of Fort Thomas WTP of Northern Kentucky water treatment plant's filter influent and effluent are given in Figure 4.7 – 4.10 during stressed and unstressed periods.

The organic material size fractions for the filter influent and effluent at the start, middle, and end of the filter run are reported in Figure 4.7 for the Fort Thomas WTP under stressed filter conditions. Organics were largely in the dissolved + fraction in the influent. However, as was observed for Milwaukee's filter effluent, the fraction of dissolved + fraction decreased and ultrafine fraction increased in the effluent during all stages of filter run. The size distribution was similar for all three stages of filter run, with dissolved + fraction at approximately 70%, ultrafine at 20% and the remaining fraction in filtrable and settleable size classes.

The distribution of various inorganic constituents into relative size fractions for the filter influent under stressed conditions are shown in Figure 4.8 for the Northern Kentucky WTP. Inorganic material principally existed in the ultrafine fraction in the filter influent, accounting for 92.3% of material. Approximately 6% of the inorganic material was equally distributed between settleable and filtrable fractions, and the remaining 1.6% was in the dissolved + fraction. The inorganic material in the filter effluent was primarily in the dissolved + fraction (approximately 95%) during all three stages of filtration. The settleable fraction was 3% at the beginning of the filter run and reduced to less than 0.5% at later stages of filter run. The dissolved + size fraction in filter influent was similarly characterized with calcium, magnesium, and silica making up the bulk of the material, with substantially smaller concentrations of iron and aluminum. The ultrafine size fraction for the filter influent was composed of approximately 98% ([C] = 33 ppm) calcium species. The filter effluent has a similar composition as the influent; however the aluminum species seem to have been almost completely removed by the filter. Most of the iron ([Fe] = 0.441 ppm) existed in the filtrable fraction in the influent and in the dissolved + fraction ([Fe] = 0.160 ppm) at all three filter stages in the effluent. Aluminum was primarily present in the ultrafine fraction ([Al] = 0.11 ppm) in the filter influent, and was evenly distributed ([Al] = 0.005 ppm) in all four fractions in the effluent. The relatively higher concentration of calcium in the ultrafine size fraction suggests that these particles resulted from calcium complexation with other particles.

The size distribution of organic material in filter influent and effluent during unstressed period is shown in Figure 4.9. The organic material in filter influent was primarily composed of dissolved + fraction (97%), as was the case for unstressed period. All the remaining material was in settleable fraction during stressed period, while for unstressed period, some material was present settleable and ultrafine fractions. In the filter effluent, bulk of the material (approximately 65%) was present in the dissolved + size range. Approximately 26% of the material was present in the ultrafine size range in the filter effluent during the stressed period, higher than that observed during unstressed period. The settleable fraction was 6 to 10%. None of the organic material was present in the filtrable fraction in the influent nor the effluent, in contrast to the unstressed period, where 3% to 7% of the material was present in this size class.

The size distribution of inorganic material present in the filter influent and effluent during unstressed period was shown in Figure 4.10. In the filter influent, most of the inorganic material was present in the dissolved + size fraction (95%), followed by settleable fraction (2.6%), ultrafine (1.8%) and filtrable fraction (0.9%). In the filter effluent, however, almost all the material was present in the dissolved fraction (greater than 99%). This is in contrast to the unstressed period, where approximately 5% of the material was present in ultrafine, filtrable and settleable fractions.

SEM images of material from the filter influent and effluent that is retained on a 0.45-μm filter pad at the Northern Kentucky's Fort Thomas water treatment facility is shown in Figure 4.11. The relative composition of the influent and effluent cakes is similar and consisted of 35% Al (Al_2O_3), 50% Si (SiO), and 10% Fe (Fe_2O_3) and 25% Al (Al_2O_3), 55% Si (SiO), and 10% Fe (Fe_2O_3), respectively based on EDAX analysis. The aggregates on the filter cake are described as spherical entities connected by bridges. TEM images of these samples also revealed circular globules with diameters around 50 nm (Figure 4.12).

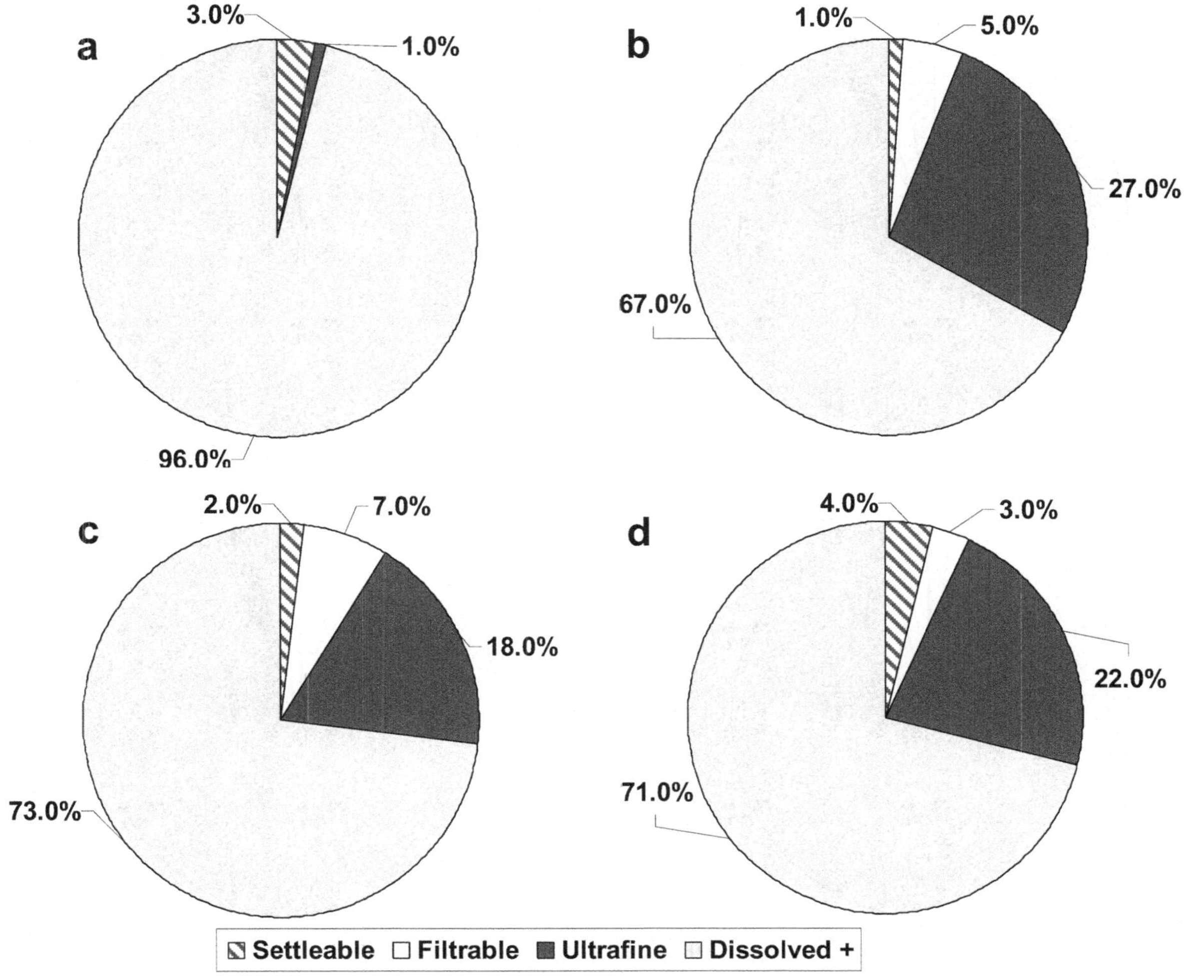

Figure 4.7 Size fractionation of the filter (a) influent and effluent TOC at the (b) start, (c) middle, and (d) end of the filter run for the Northern Kentucky water treatment facility during the stressed sampling period (influent $[TOC]_T$ = 1.78 ppm, effluent $[TOC]_T$ = 1.62 ppm)

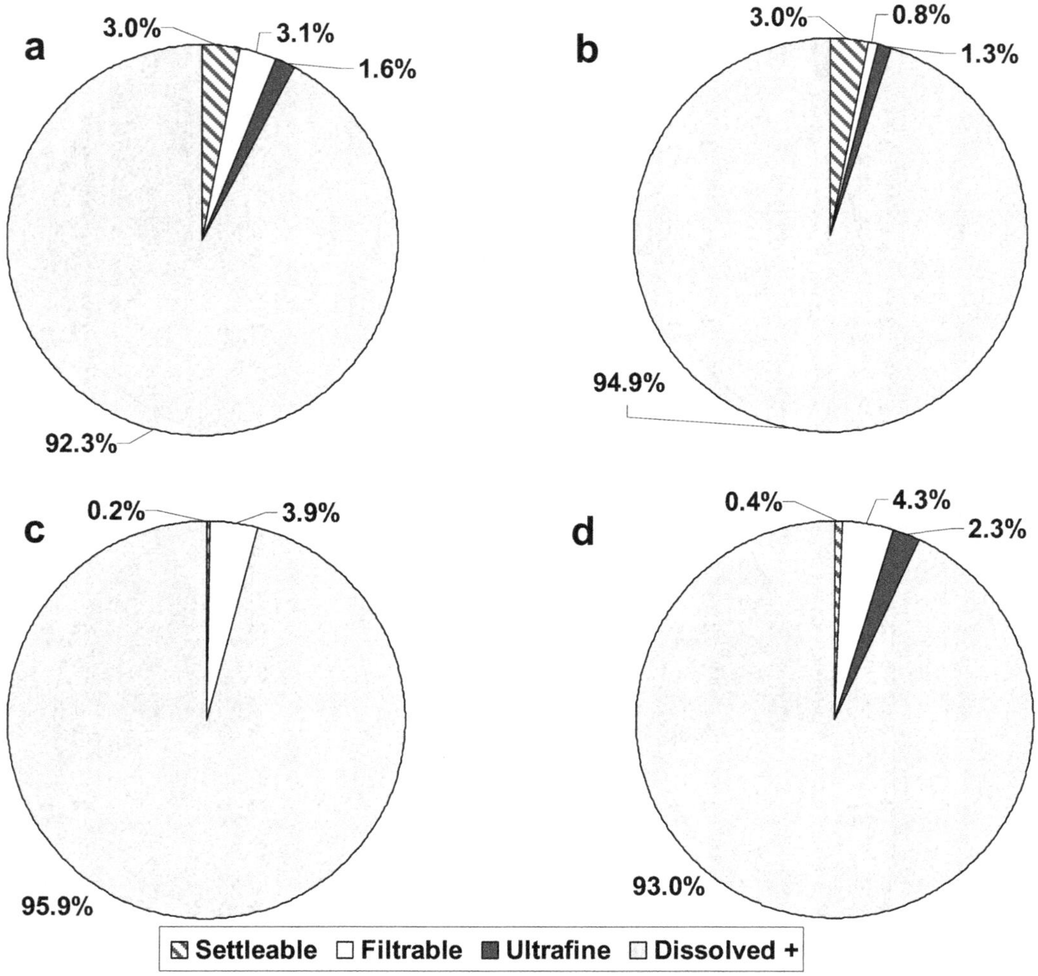

Figure 4.8 Size fractionation of the filter (a) influent and effluent elemental composition at the (b) start, (c) middle, and (d) end of the filter run for the Northern Kentucky water treatment facility during the stressed sampling period (influent $[Metals]_T$ = 52.51 ppm, effluent $[Metals]_T$ = 49.57 ppm)

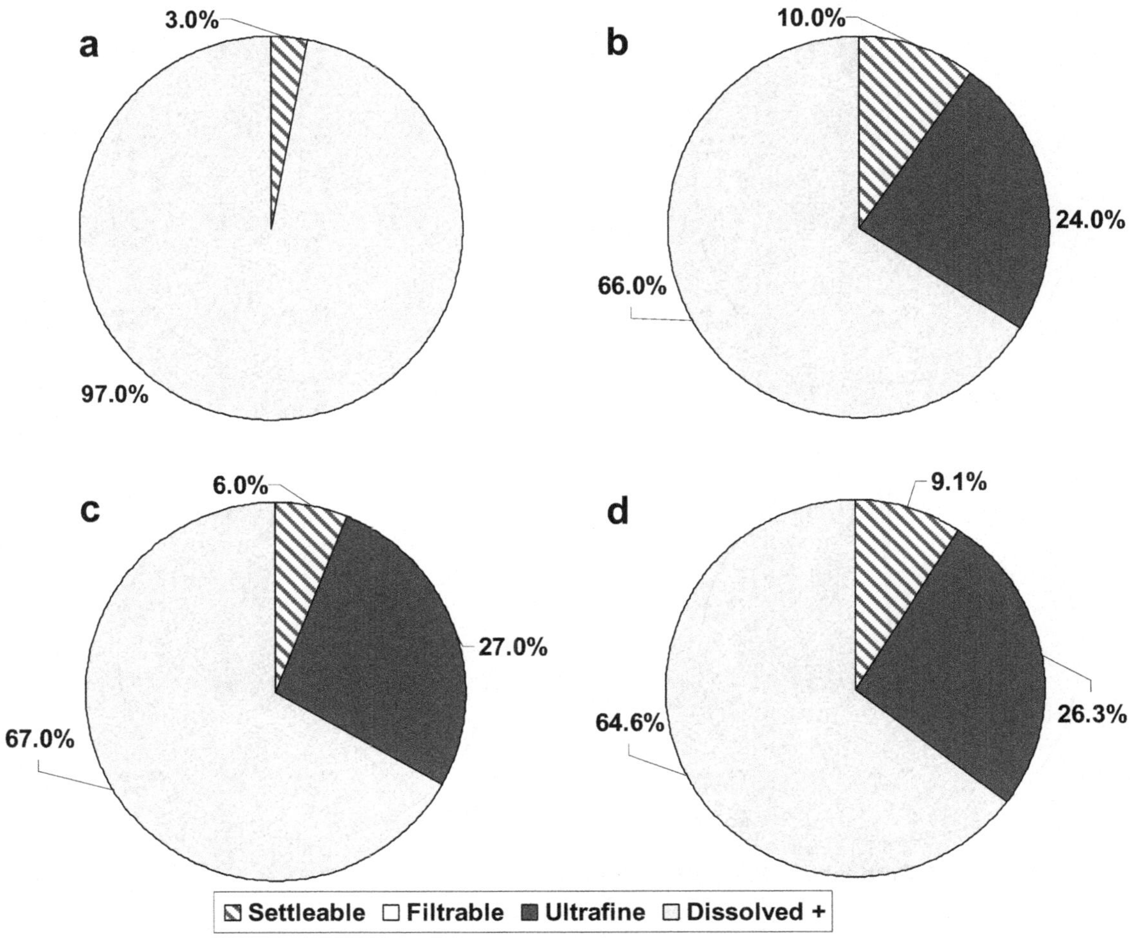

Figure 4.9 Size fractionation of the filter (a) influent and effluent TOC at the (b) start, (c) middle, and (d) end of the filter run for the Northern Kentucky water treatment facility during the unstressed sampling period (influent $[TOC]_T = 1.80$ ppm, effluent $[TOC]_T = 1.68$ ppm)

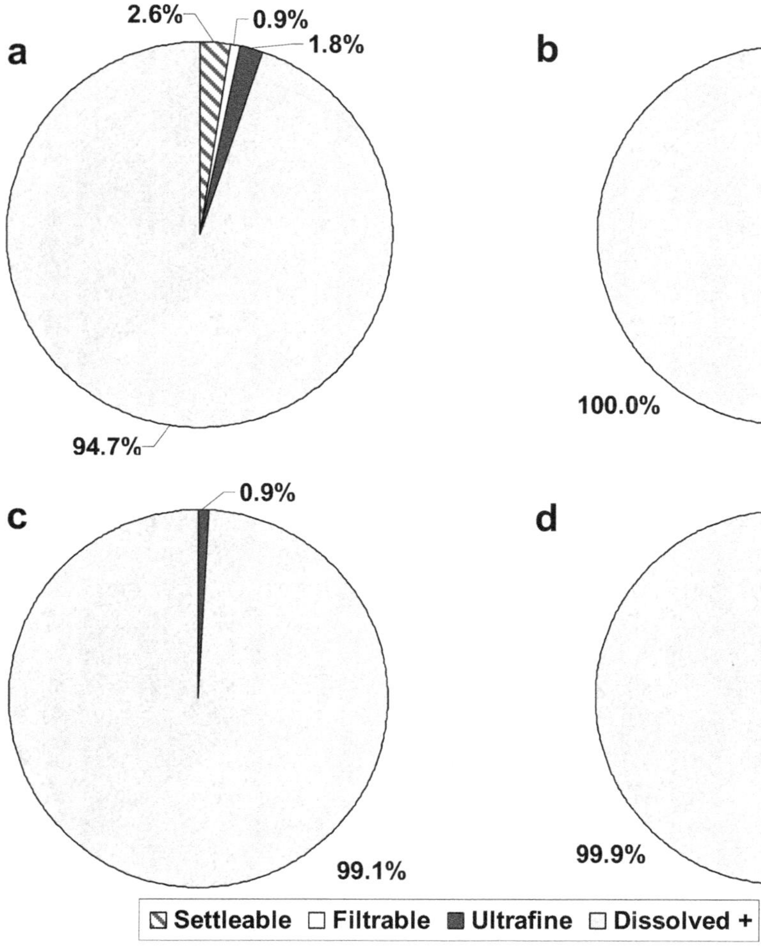

Figure 4.10 Size fractionation of the filter (a) influent and effluent elemental composition at the (b) start, (c) middle, and (d) end of the filter run for the Northern Kentucky water treatment facility during the unstressed sampling period (influent $[Metals]_T$ = 41.11 ppm, effluent $[Metals]_T$ = 39.98 ppm)

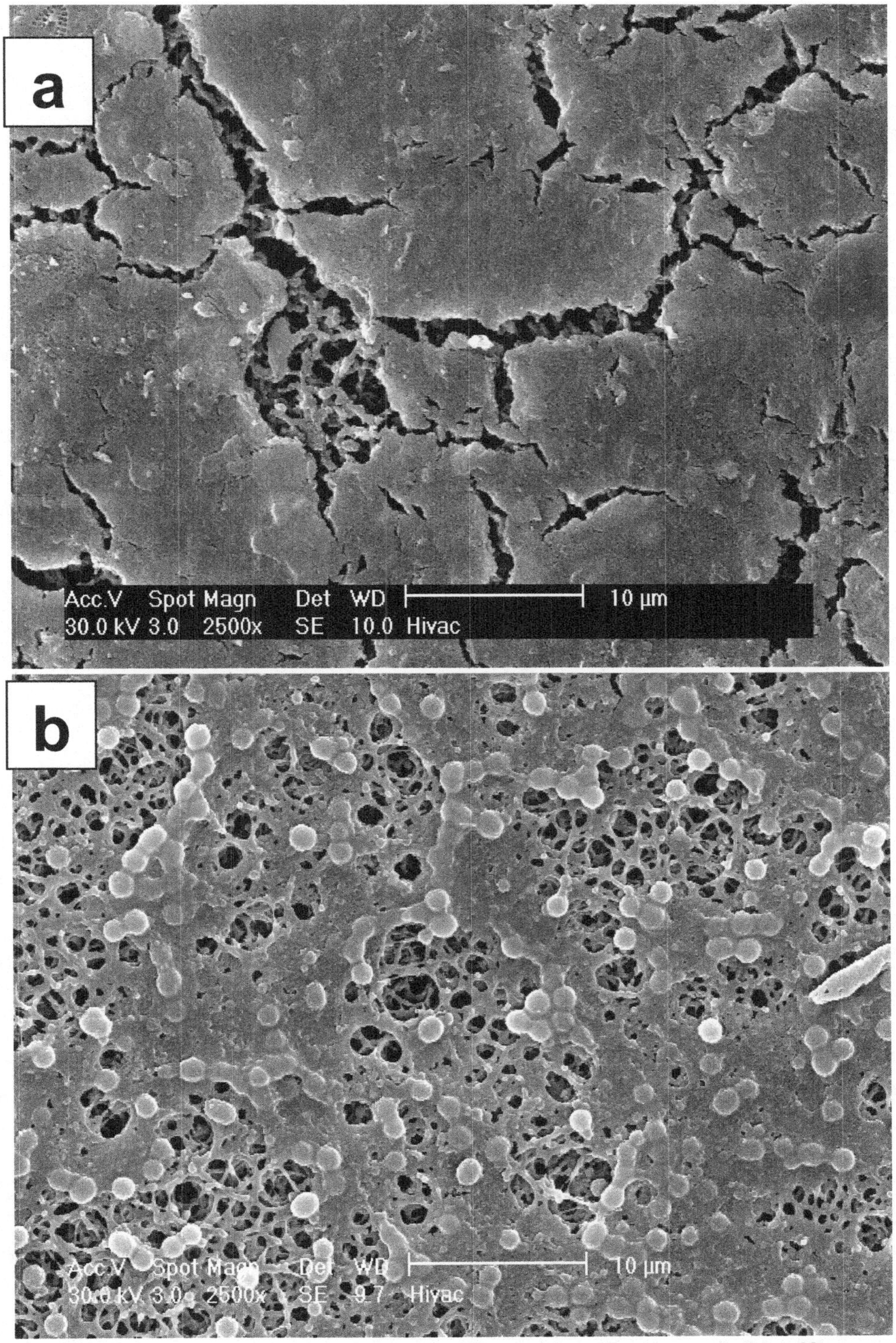

Figure 4.11 SEM images of material from the filter (a) influent and (b) effluent retained on a 0.45-µm filter pad at the Northern Kentucky water treatment facility during the unstressed operating period. The effluent sample was collected from Filter #4 at the end of the filter run. The composition the influent and effluent cakes is 35% Al (Al_2O_3), 50% Si (SiO), and 10% Fe (Fe_2O_3) and 25% Al (Al_2O_3), 55% Si (SiO), and 10% Fe (Fe_2O_3), respectively based on EDAX analysis

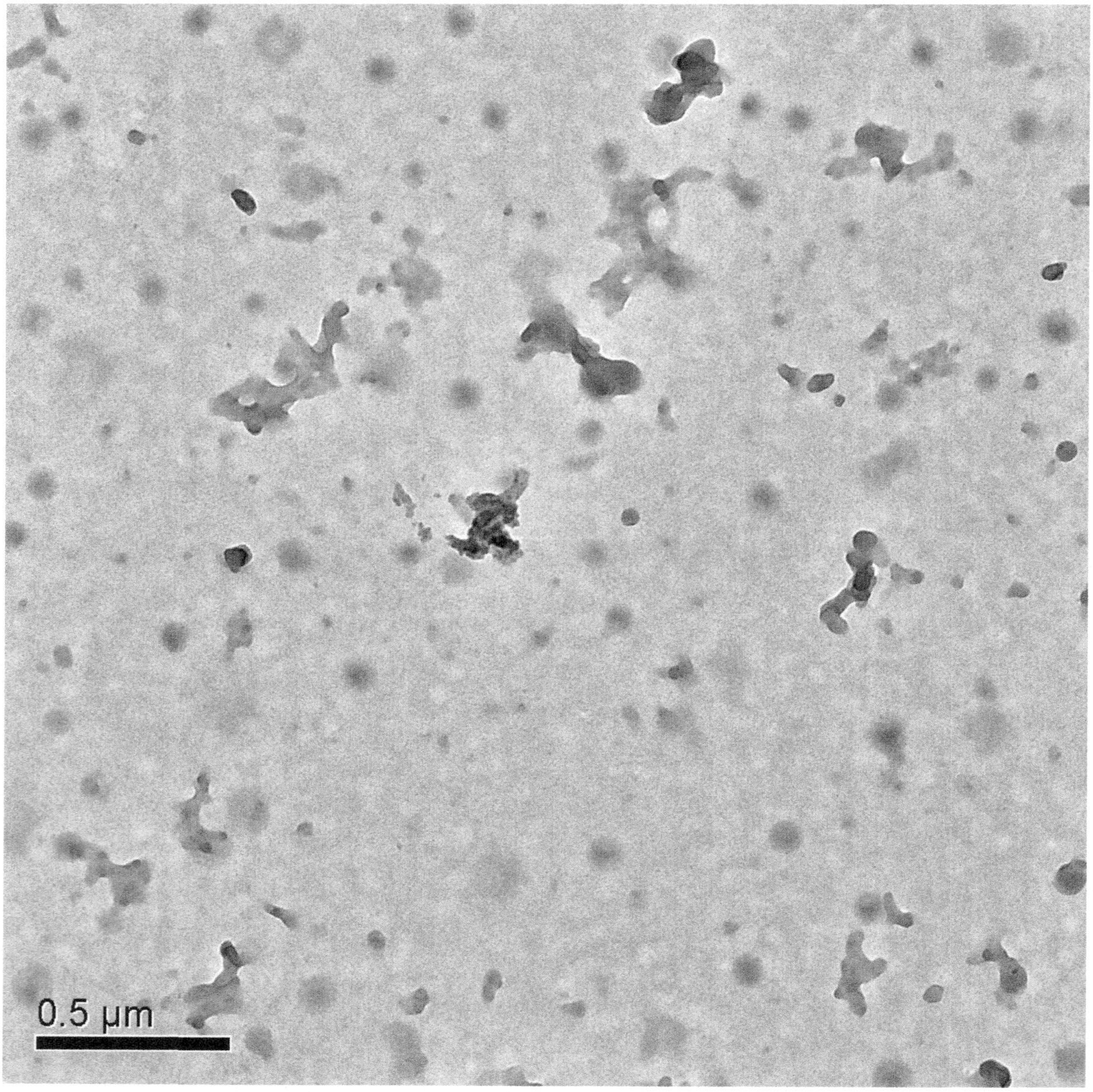

Figure 4.12 TEM image of ultrafine particles deposited on a copper mesh microgrid from the filter effluent during the unstressed operating period at the Northern Kentucky water treatment facility. The effluent sample was collected at Filter #10 at the start of the filter run. The small spherical globules have a diameter of approximately 50 nm

Buffalo Pound Water Administration

Mass size distributions of organic and inorganic material present in the Buffalo Pound water treatment plant's filter influent and effluent during stressed period are shown in Figures 4.13 – 4.14.

Significant amount of organic material mass in the filter influent was distributed between dissolved +, ultrafine and settleable classes (Figure 4.13a). A small fraction of organic material was present in the filtrable fraction (1.6%). In the filter effluent, the settleable fraction decreased and the dissolved + and ultrafine fractions increased (Figure 4.13b-d). No organic material was present in the filtrable fraction at the beginning and end of filter run, and a small fraction (1.3%) was present in the filter effluent at the middle of the run. The ultrafine fraction in the filter effluent gradually increased during the course of filter run, while the settleable fraction decreased.

The mass size distribution of inorganic material is shown in Figure 4.14. Most of the inorganic material was present in the dissolved + and ultrafine fractions, at 58% and 41% respectively. In the filter effluent, the distribution was similar, with approximately 56% to 65% in the dissolved + size range, and 30 to 38% in the ultrafine range. The mass fraction in the dissolved + and ultrafine was similar at the beginning and the end of the filter run. During the middle of the run, dissolved + fraction decreased with similar increase in ultrafine fraction.

The SEM images of influent and effluent material retained on a 0.45-μm filter pad at the Buffalo Pound facility is shown in Figure 4.15. The composition of the influent cake was principally (22%) aluminum (Al_2O_3) and organic materials. The effluent cake was also composed of 5% Al (Al_2O_3) and organic materials based on EDAX analysis.

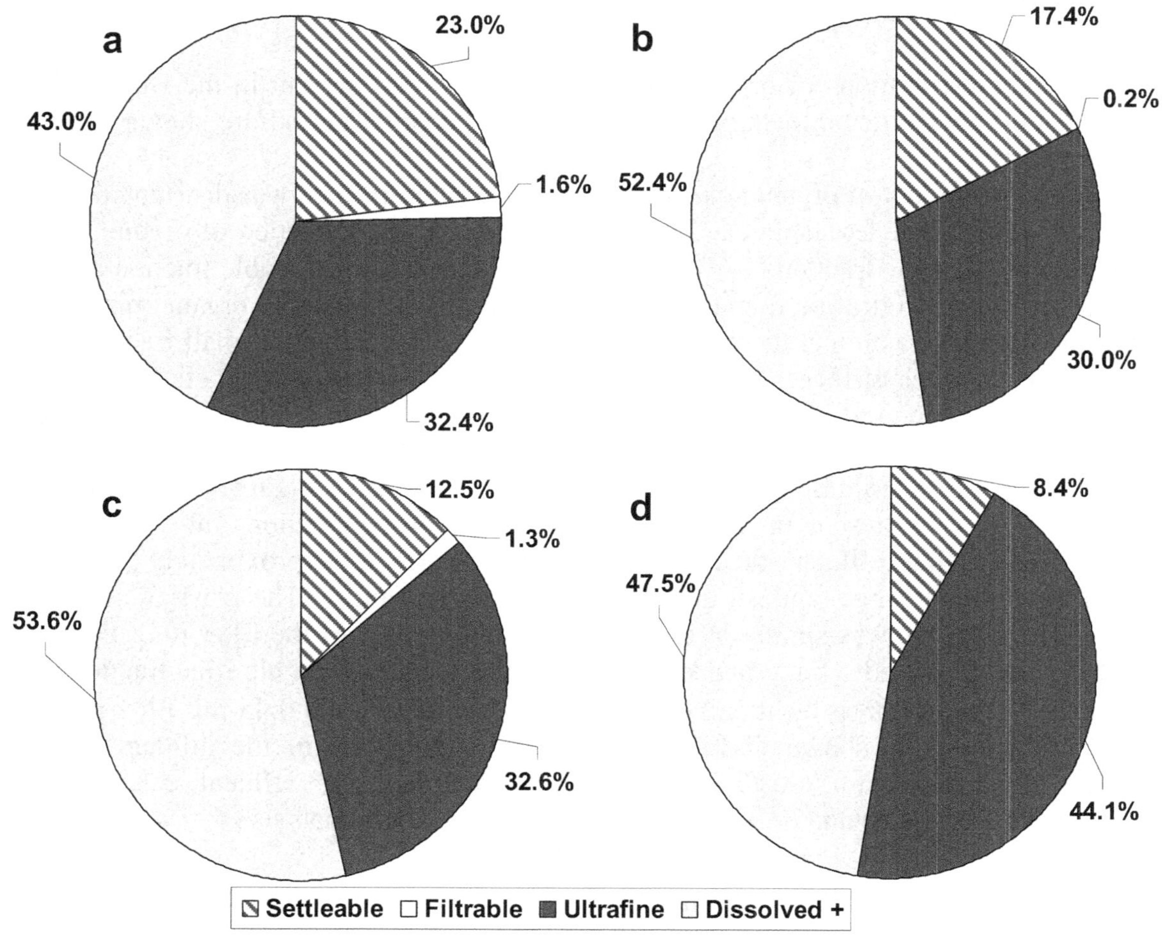

Figure 4.13 Size fractionation of the filter (a) influent and effluent TOC at the (b) start, (c) middle, and (d) end of the filter run for the Buffalo Pound water treatment facility during the stressed sampling period (influent $[TOC]_T$ = 6.49 ppm, effluent $[TOC]_T$ = 5.57 ppm)

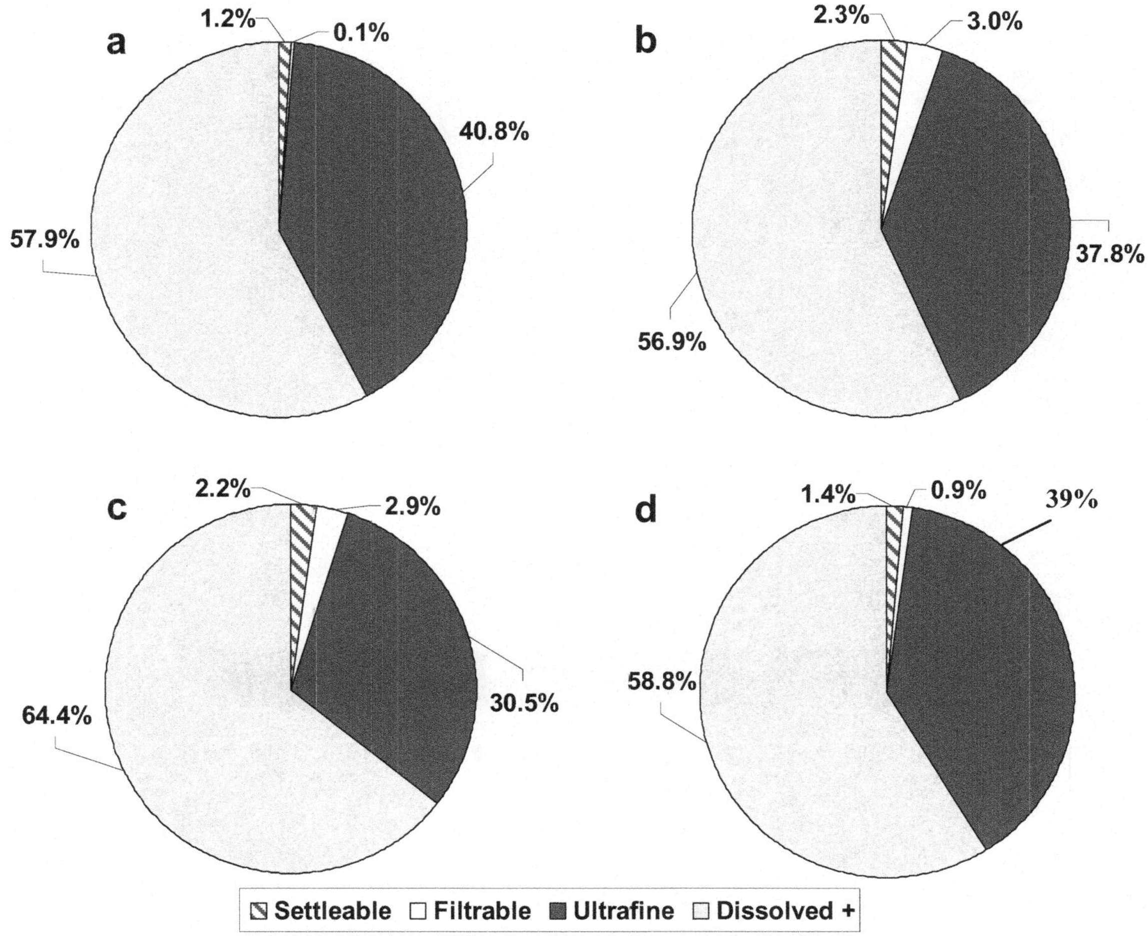

Figure 4.14 Size fractionation of the filter (a) influent and effluent elemental composition at the (b) start, (c) middle, and (d) end of the filter run for the Buffalo Pound water treatment facility during the stressed sampling period (influent $[Metals]_T$ = 76.74 ppm, effluent $[Metals]_T$ = 78.68 ppm)

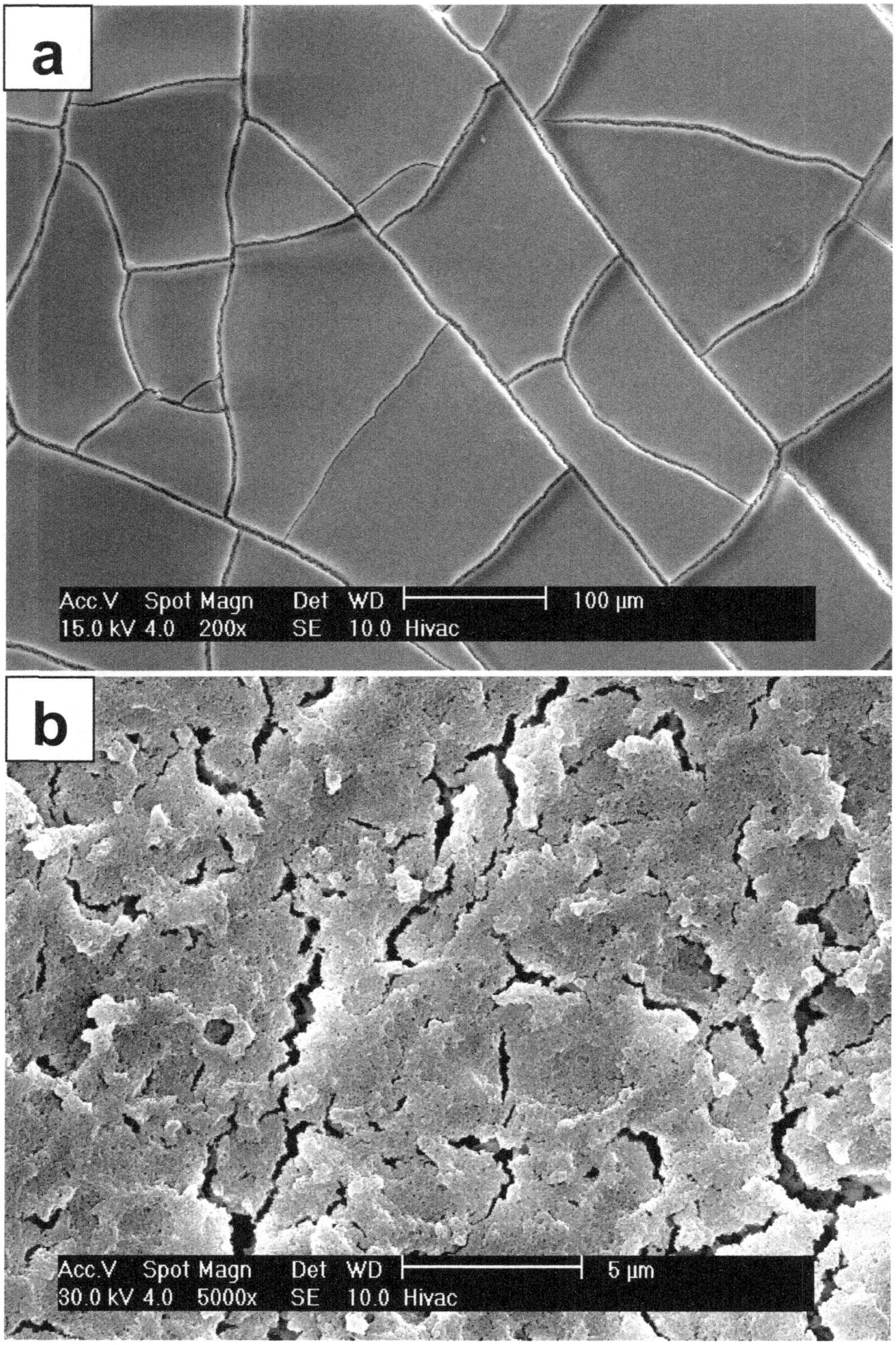

Figure 4.15 SEM images of material from the filter (a) influent and (b) effluent retained on a 0.45-µm filter pad at the Buffalo Pound water treatment facility. The composition of the (a) influent cake is principally (22%) aluminum (Al_2O_3) and organic materials. The effluent cake is composed of 5% Al (Al_2O_3) and organic materials based on EDAX analysis

City of Cincinnati – Department of Water Works

Mass size distributions of organic and inorganic material present in the City of Cincinnati's water treatment plant's filter influent and effluent are given in Figures 4.16 – 4.19 during stressed and unstressed periods.

The mass size distribution of organic material present in the filter influent and effluent are presented in Figure 4.16 during the stressed period. Most of the organic material in the filter influent is present in the dissolved + size range (88.9%), followed by filtrable fraction (9.1%) and 1% in each of settleable and ultrafine fractions. In the filter effluent, the dissolved + size fraction increased from 71% at the beginning of the run, to 92% during the middle of the run and subsequently reduced to 86% at the end of the run. The settleable fraction was higher at the beginning of the run (21%) and reduced to 7% during the middle and end of run. The filtrable fraction was 7% at the beginning of the run, 1% at the middle of the run and was 4% at the end of the run. A small amount of organic material was present during the beginning stages of the filter run (1%), none at the middle of the run and eventually increased to 3% at the end of the run. About 20% of the TOC was removed during filtration.

The mass distribution of inorganic material in the filter influent and effluent during stressed period is shown in Figure 4.17. Most of the inorganic material in the filter influent and effluent was in dissolved + size fraction. The settleable fraction was relatively high in the filter effluent during the beginning of the run and gradually decreased during the course of filter run. The influent is comprised mostly of calcium, magnesium, and silica, which together account for 99.8% of the total mass of metal species. Aluminum made up just 0.2% of the total mass fraction. Manganese was also present in the filter influent ([Mn] = 0.015 ppm), though at a substantially lower concentration than the other metals.

The mass distribution of organic matter in the filter influent and effluent during the unstressed period is shown in Figure 4.18. The size distribution was significantly different than the stressed period. While during stressed period, the bulk of the organic material was present in the dissolved + size class, during the unstressed period, the bulk of the organic matter was present in dissolved + and ultrafine fraction. Approximately 59% of the organic material in the filter influent was present in the dissolved + size class, followed by 29% in the ultrafine size class and 12% equally distributed between settleable and filtrable size classes. The dissolved + size fraction of organic material in the filter effluent decreased from 57% at the beginning of the filter run to 47.5% at the end of the run. The ultrafine fraction was 30% at the beginning of the run, and decreased to 26.3% during the middle of the run. Towards the end of the filter run, the ultrafine fraction increased to 43.6%. The settleable fraction was initially higher at 13% at the beginning of the run and decreased to between 6% and 9% during the later stages of filter run.

The mass distribution of inorganic material in the filter influent and effluent during the unstressed sampling period is shown in Figure 4.19. Similar to organic material, the size distribution of inorganic material also differed significantly during the stressed and unstressed periods. During the stressed period, more than 98% of the material was in the dissolved + size class. However, for the unstressed period, a significant amount of inorganic material was present in the dissolved + and ultrafine size classes. The size distribution of inorganic material was similar at the beginning and end of filter run, with approximately 58% in the dissolved + size range and 39% in the ultrafine range. At the middle of the filter run, the dissolved + fraction was higher at 64% and the ultrafine fraction was lower at 31%. The filtrable fraction was approximately 3% at the beginning and middle of filter run and was below 1% at the end of the run. The settleable fraction was approximately 2.3% during the beginning and middle of filter run and was 1.4% at the end of the run.

SEM images of the material present in the influent and effluent retained on a 0.45-µm filter pad at the Cincinnati Water Works water treatment facility are shown in Figures 4.20 and 4.21 respectively. The composition of the influent cake was 27% Al (Al_2O_3), 48% Si (SiO), and 15% Fe (Fe_2O_3) based on EDAX analysis. The biota shown in Figures 4.20a and 4.20b are diatoms Melosira and Cylotella respectively. The particle in Figure 4.20d was mostly composed of silica (SiO) (95%). The effluent sample shown in Figure 4.21 was collected from Filter #7 during the middle of the filter run during the stressed period. The composition the effluent cake shown in Figure 4.21a was 35% Al (Al_2O_3), and 50% Si (SiO) based on EDAX analysis. The diatom shown in Figure 4.21b is Cyclotella and the biota in Figure 4.21c and 4.21d are Streptococcus. Druing stressed periods, algal content in the raw water is higher than during non-stressed periods. TEM images of the filter effluent are shown in Figure 4.22. The spherical particle diameters were approximately 50 nm.

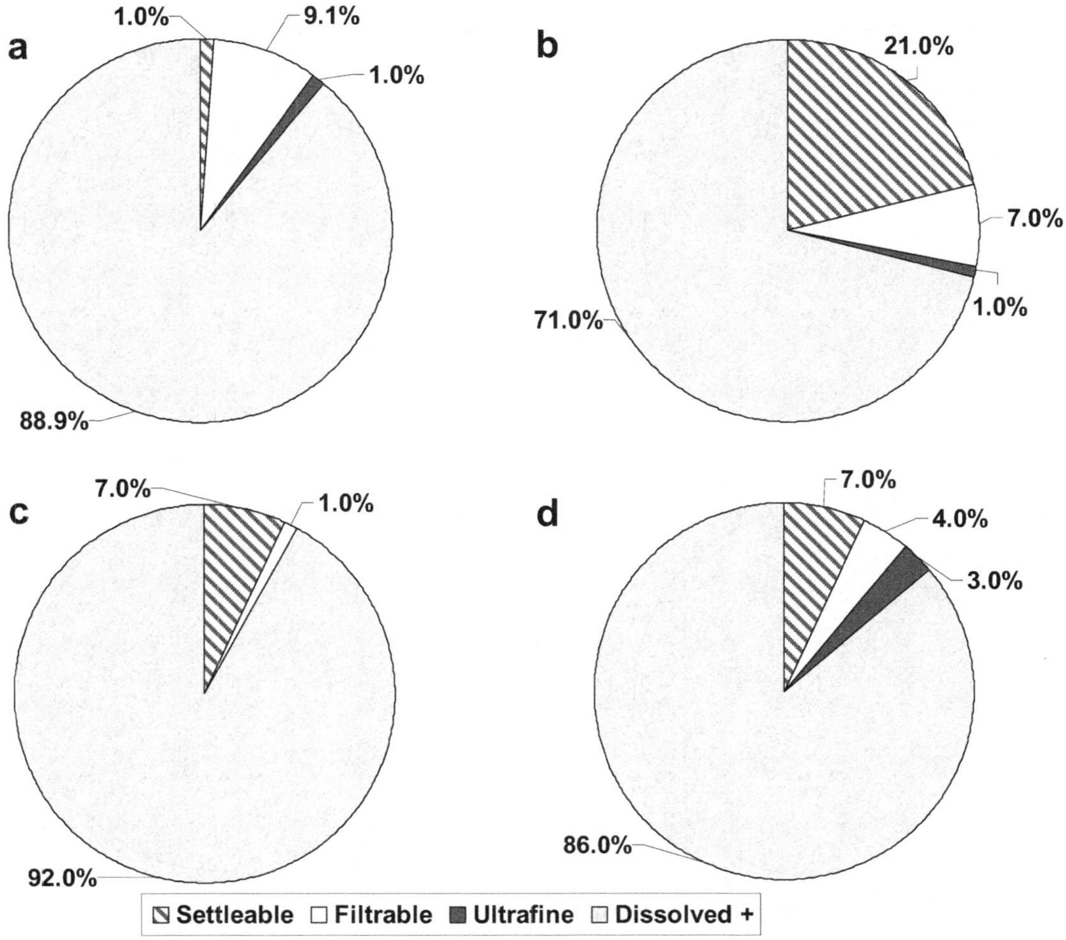

Figure 4.16 Size fractionation of the filter (a) influent and effluent TOC at the (b) start, (c) middle, and (d) end of the filter run for the Cincinnati water treatment facility during the stressed sampling period (influent $[TOC]_T = 3.3$ ppm, effluent $[TOC]_T = 2.2$ ppm)

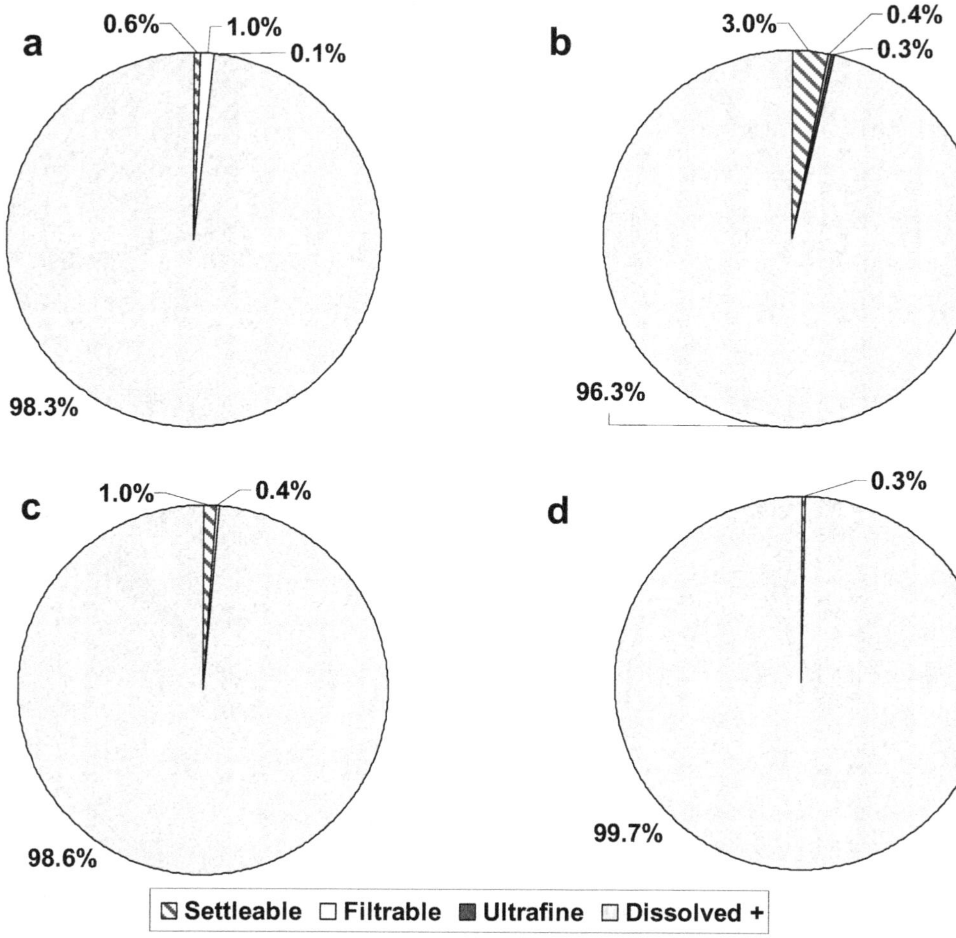

Figure 4.17 Size fractionation of the filter (a) influent and effluent elemental composition at the (b) start, (c) middle, and (d) end of the filter run for the Cincinnati water treatment facility during the stressed sampling period (influent $[Metals]_T = 52.5$ ppm, effluent $[Metals]_T = 53$ ppm)

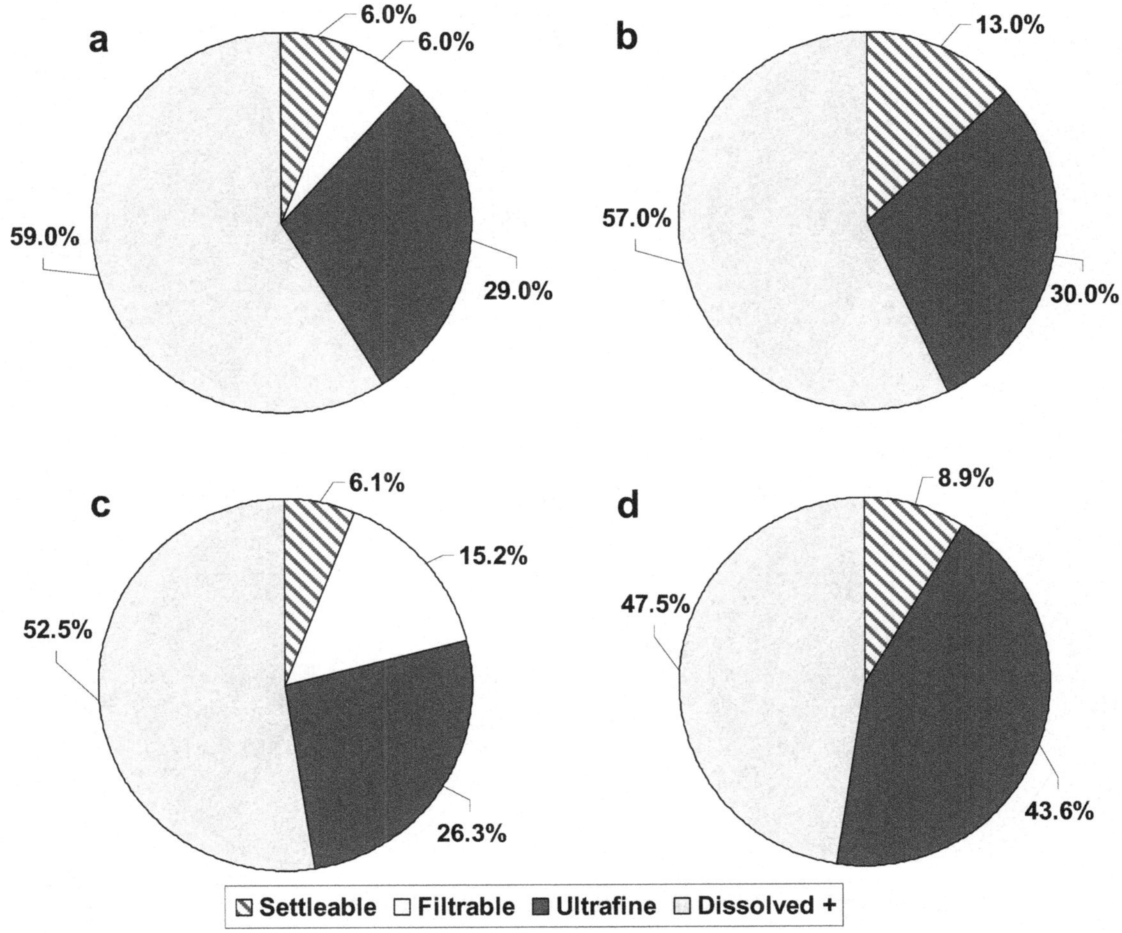

Figure 4.18 Size fractionation of the filter (a) influent and effluent TOC at the (b) start, (c) middle, and (d) end of the filter run for the Cincinnati water treatment facility during the unstressed sampling period (influent $[TOC]_T = 2.25$ ppm, effluent $[TOC]_T = 2.04$ ppm)

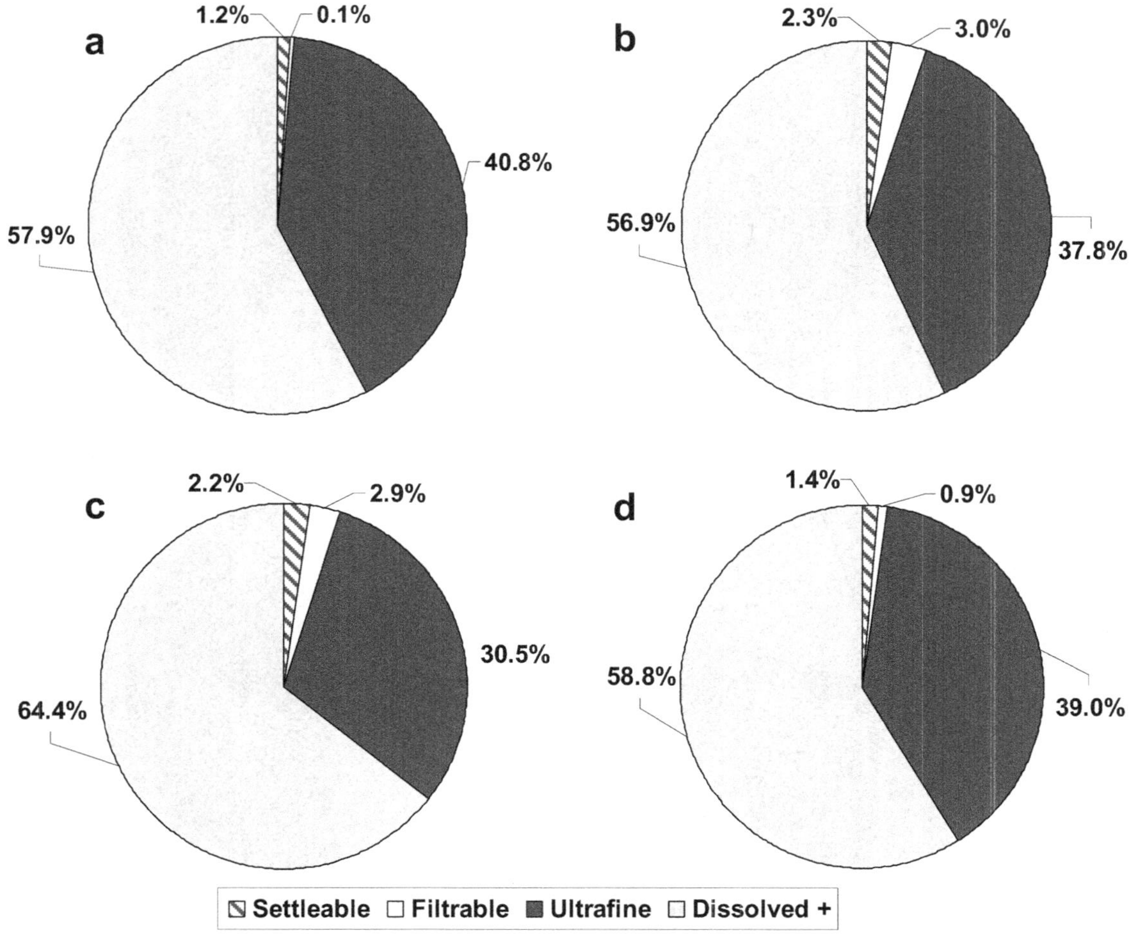

Figure 4.19 Size fractionation of the filter (a) influent and effluent elemental composition at the (b) start, (c) middle, and (d) end of the filter run for the Cincinnati water treatment facility during the unstressed sampling period (influent $[Metals]_T$ = 52.6 ppm, effluent $[Metals]_T$ = 53 ppm)

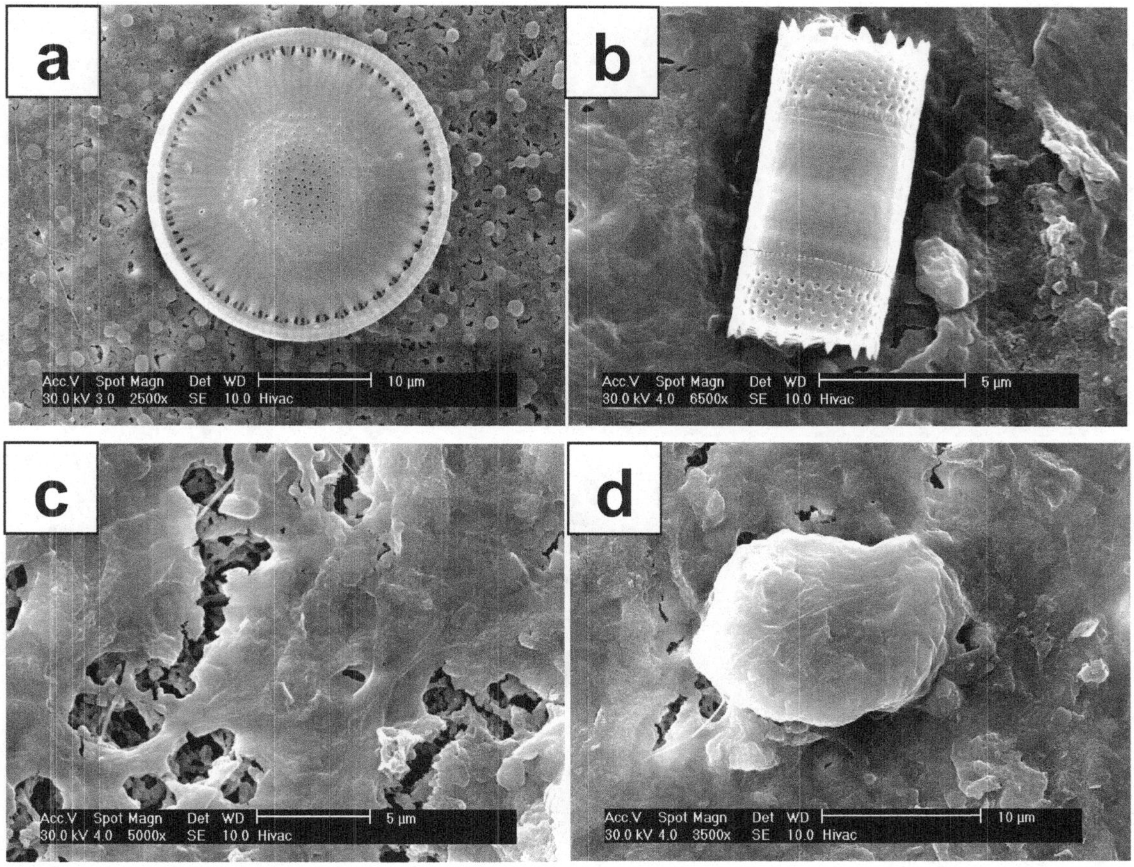

Figure 4.20 SEM images of material from the filter influent retained on a 0.45-µm filter pad at the City of Cincinnati water treatment facility during the stressed operating period. The biota were identified as (a) *Melosira* and (b) *Cyclotella*. The composition of the (c) influent cake is 27% Al (Al_2O_3), 48% Si (SiO), and 15% Fe (Fe_2O_3) based on EDAX analysis. The particle in (d) is mostly (95%) composed of silica (SiO)

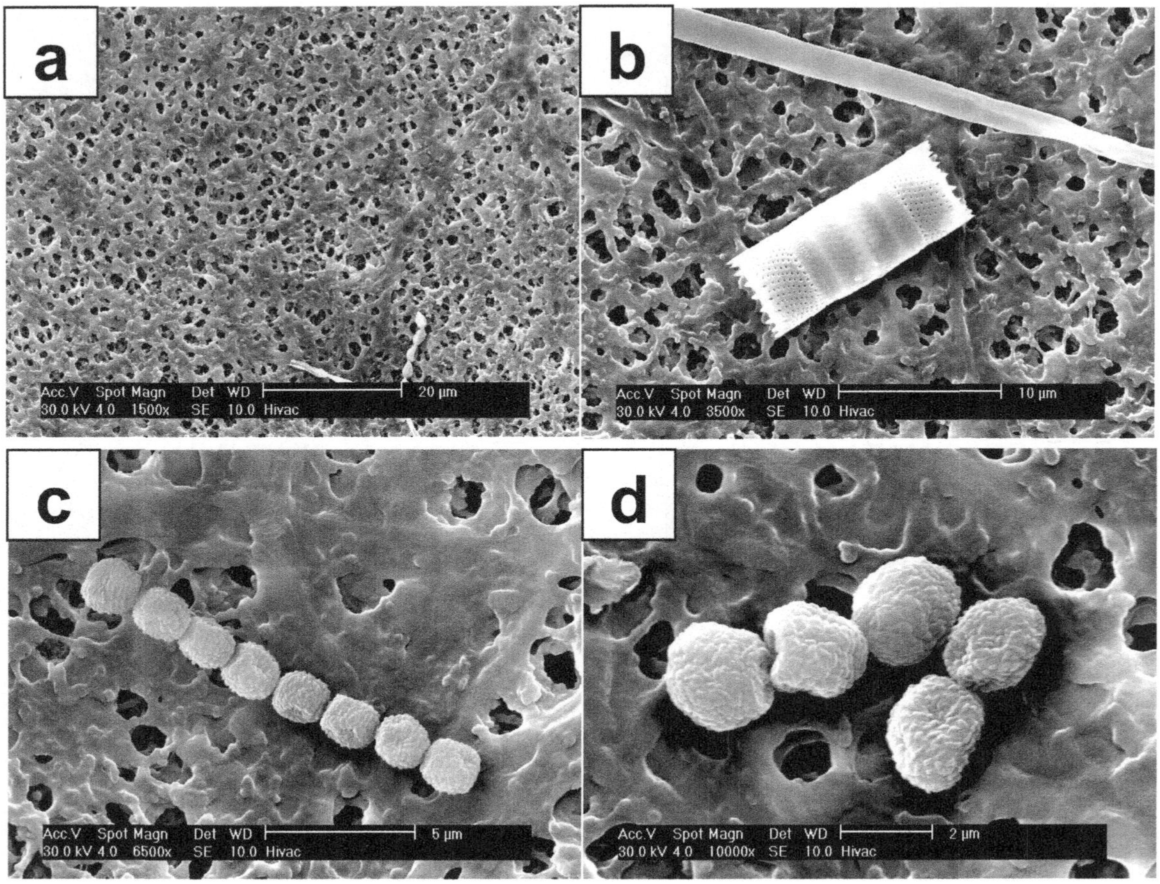

Figure 4.21 SEM images of material from the filter effluent retained on a 0.45-µm filter pad at the City of Cincinnati water treatment facility during the stressed operating period showing a (b) *Cyclotella* and (c - d) *Streptococcus*. The effluent sample was collected from Filter #7 during the middle of the filter run. The composition of the effluent cake (a) is 35% Al (Al_2O_3), 50% Si (SiO) based on EDAX analysis

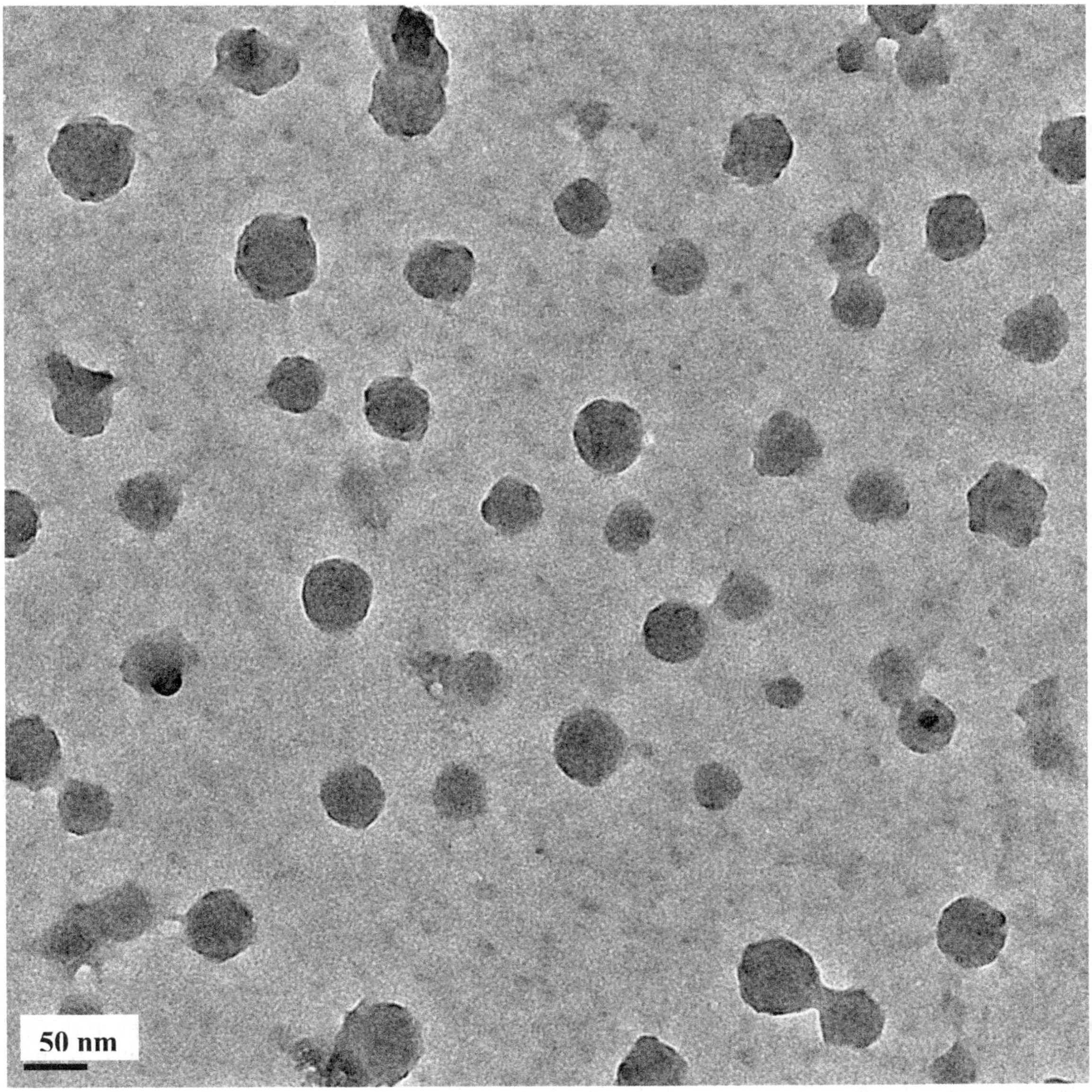

Figure 4.22 TEM image of ultrafine particles deposited on a copper mesh microgrid from the filter influent during the stressed operating period at the City of Cincinnati water treatment facility. The small spherical globules have a diameter of approximately 50 nm

City of Houston – East Water Purification Plant (EWPP)

The mass distributions of organic and inorganic material in the influent and effluent of City of Houston's East Water Purification Plant's filters are shown in Figures 4.23 – 4.26. The distribution of organic material in various size classes for the filter influent and effluent during the stressed filter period are shown in Figure 4.23. Most of the organic material in the filter influent is in the dissolved + size range. However, in the filter effluent, during the beginning of the filter run, most of the organic material was present in the dissolved + and settleable size fractions (approximately 57% and 40%, respectively). The dissolved + size fraction increased to 73.3% at the middle of the filter run and to 87% at the end of the run. Correspondingly, the settleable fraction decreased from 39% at the beginning of the filter run to 21% at the middle of the run to 6% at the end of the run. The filtrable fraction of organic material increased from 1% at the beginning of the run to 4% at the middle and end of run. The ultrafine fraction of organic material essentially remained within 2 to 3%.

The distribution of inorganic material into relative size fractions for the filter influent and effluent under stressed conditions are reported in Figure 4.24. Inorganic material in the filter influent was primarily in the dissolved + size fraction (approximately 94%), followed by the filtrable (3.1%), settleable (1.8%) and ultrafine fraction (1%). The inorganic material in the filter effluent was also primarily composed of the dissolved + size fraction. However, the ultrafine fraction of inorganic material was higher in the filter effluent, at approximately 9% during the beginning and middle of filter run and approximately 6% at the end of the filter run. The settleable and filterable fractions in the effluent were lower than influent. The inorganic material in the influent and effluent was primarily composed of silica, calcium, and magnesium; aluminum was also present but at substantially lower concentrations. The dissolved + size fraction, which constituted the largest fraction in all of the water samples, was composed of silica, calcium, and magnesium, in addition to relatively low concentrations of aluminum. Aluminum in the influent, was primarily in the ultrafine ([Al] = 0.633 ppm) and dissolved + ([Al] = 0.119 ppm) size fractions.

The mass distribution of organic material in the filter influent and effluent during unstressed period is shown in Figure 4.25. The organic material in the filter influent was present in all size classes. The dissolved + size class has the highest percentage, at 54%, followed by ultrafine at 25%, filterable at 15% and settleable at 6%. Most of the organic material in the filter effluent at the beginning of the run was present in the dissolved + size range (76%), followed by ultrafine (16%), filterable (6%) and settleable (2%). At the middle of the run, the dissolved + size fraction decreased to 62%, while the ultrafine fraction increased to 31%. None of the organic material was present in the filtrable fraction and 7% was present in the settleable fraction. The distribution at the end of the filter run was similar to that at the middle of filter run.

The mass distribution of inorganic material in the filter influent and effluent during the unstressed period is shown in Figure 4.26. In contrast to the size distribution during the stressed period where most of the inorganic material was present in the dissolved + size class, during the unstressed period, the inorganic material was present primarily in the dissolved + (61%) and ultrafine fractions (38%). The size distribution in the filter effluent was similar, with most of the material present in these two size classes. At the beginning of the run, a small percentage of material was present in the settleable size class and during the middle of the run, no material was present in the settleable class, and a similar amount was present in filtrable class.

SEM images of the material in the filter influent and effluent retained on a 0.45-μm filter pad at the Houston water treatment facility are shown in Figure 4.27. The composition the influent and effluent cakes was similar: 13% Al (Al_2O_3) and 8% Si (SiO) based on EDAX analysis, with the remaining material being of organic origin. In addition, a TEM image of the filter effluent is given in Figure 4.28.

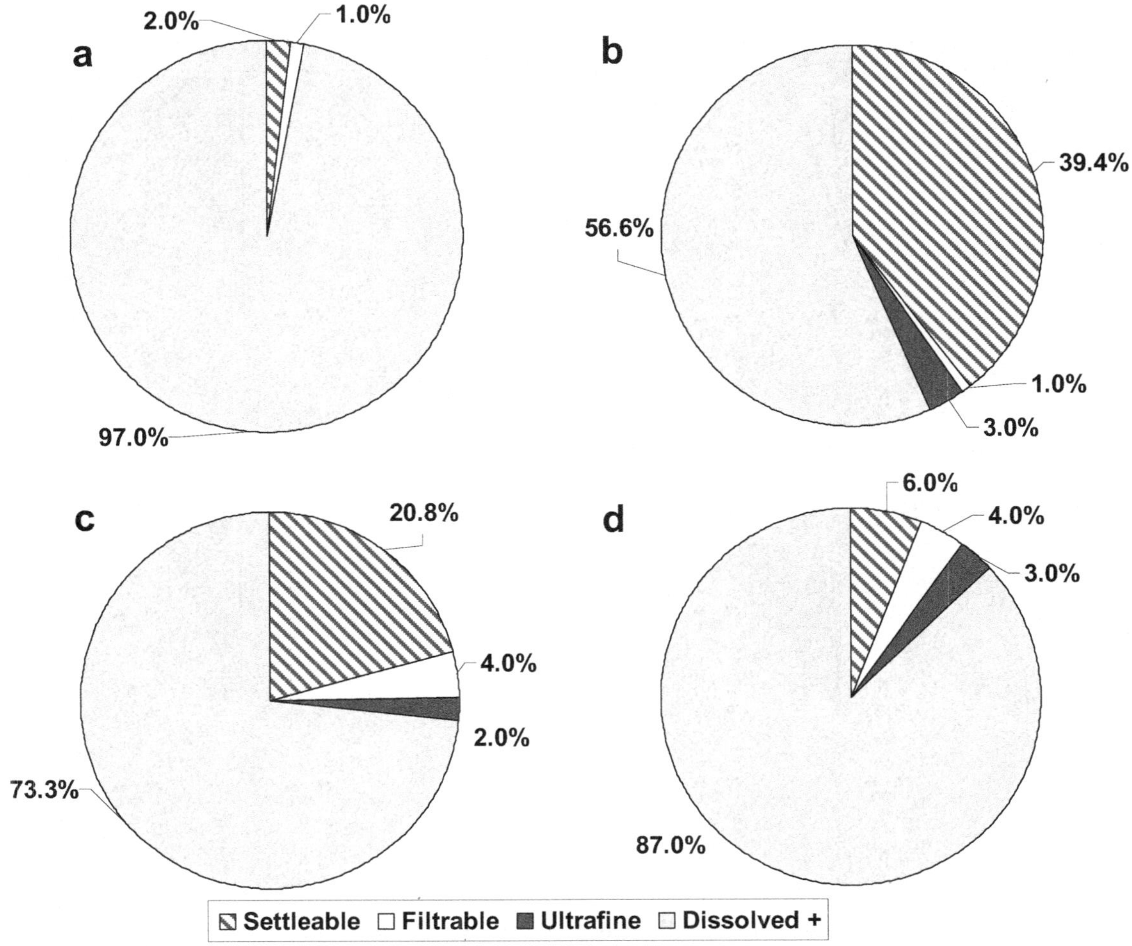

Figure 4.23 Size fractionation of the filter (a) influent and effluent TOC at the (b) start, (c) middle, and (d) end of the filter run for the Houston water treatment facility during the stressed sampling period (influent $[TOC]_T = 6.49$ ppm, effluent $[TOC]_T = 5.57$ ppm)

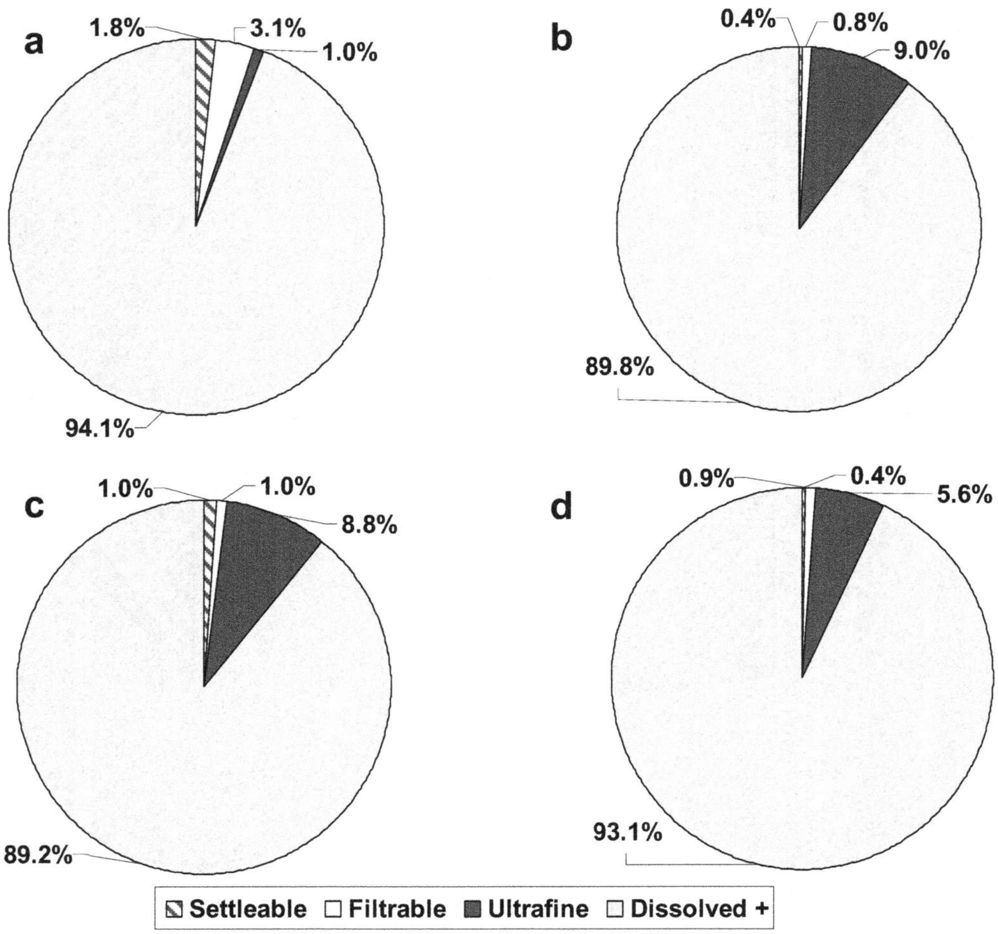

Figure 4.24 Size fractionation of the filter (a) influent and effluent elemental composition at the (b) start, (c) middle, and (d) end of the filter run for the Houston water treatment facility during the stressed sampling period (influent $[Metals]_T$ = 57.89 ppm, effluent $[Metals]_T$ = 55.01 ppm)

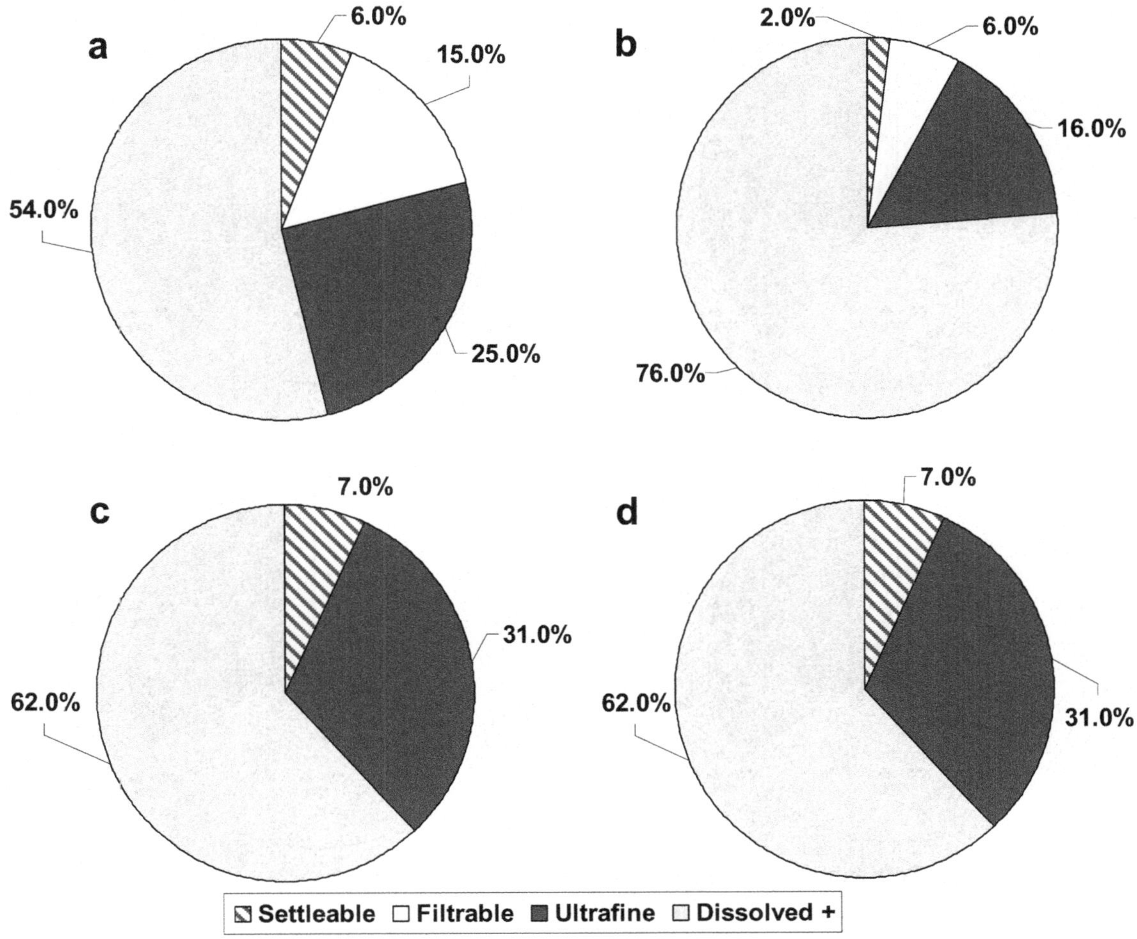

Figure 4.25 Size fractionation of the filter (a) influent and effluent TOC at the (b) start, (c) middle, and (d) end of the filter run for the Houston water treatment facility during the unstressed sampling period (influent $[TOC]_T = 2.25$ ppm, effluent $[TOC]_T = 2.04$ ppm)

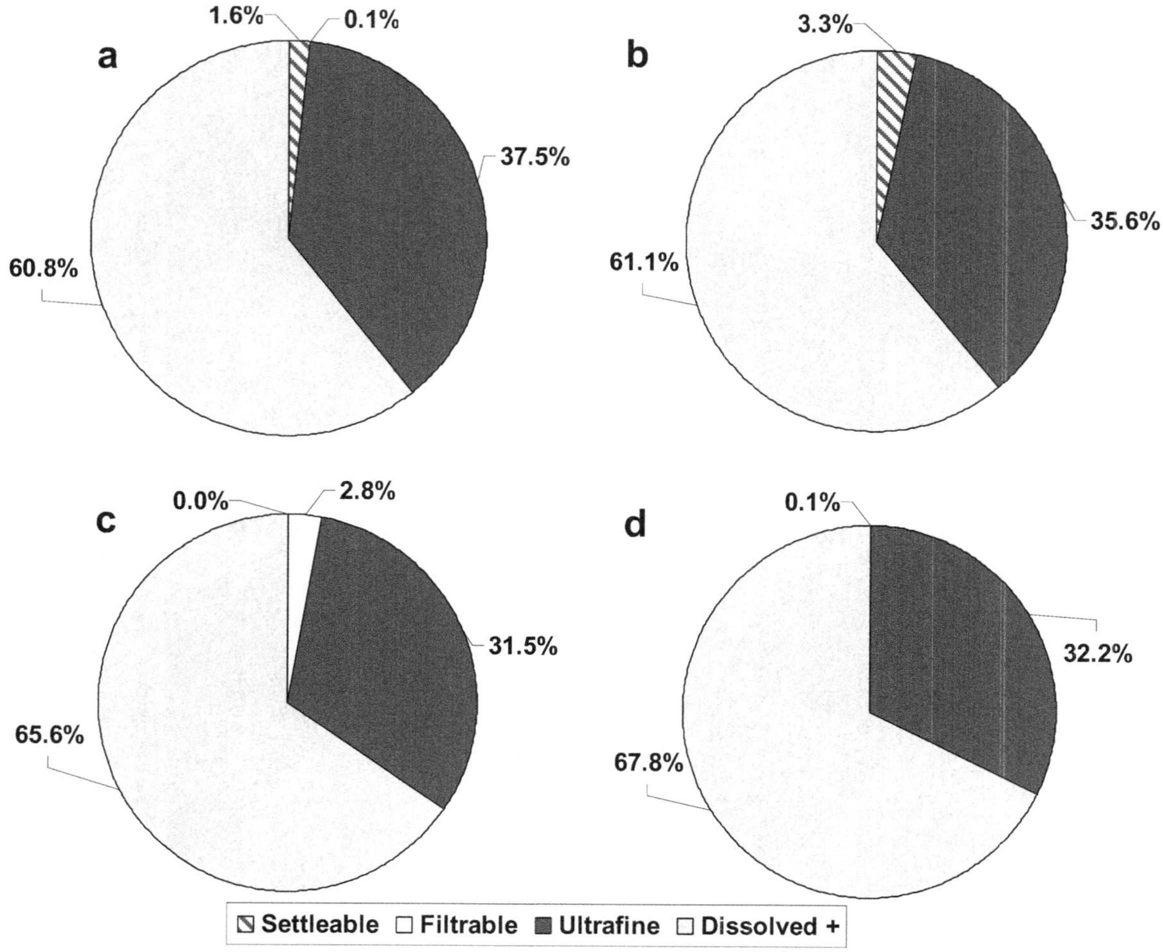

Figure 4.26 Size fractionation of the filter (a) influent and effluent elemental composition at the (b) start, (c) middle, and (d) end of the filter run for the Houston water treatment facility during the unstressed sampling period (influent $[Metals]_T$ = 64.89 ppm, effluent $[Metals]_T$ = 59.70 ppm)

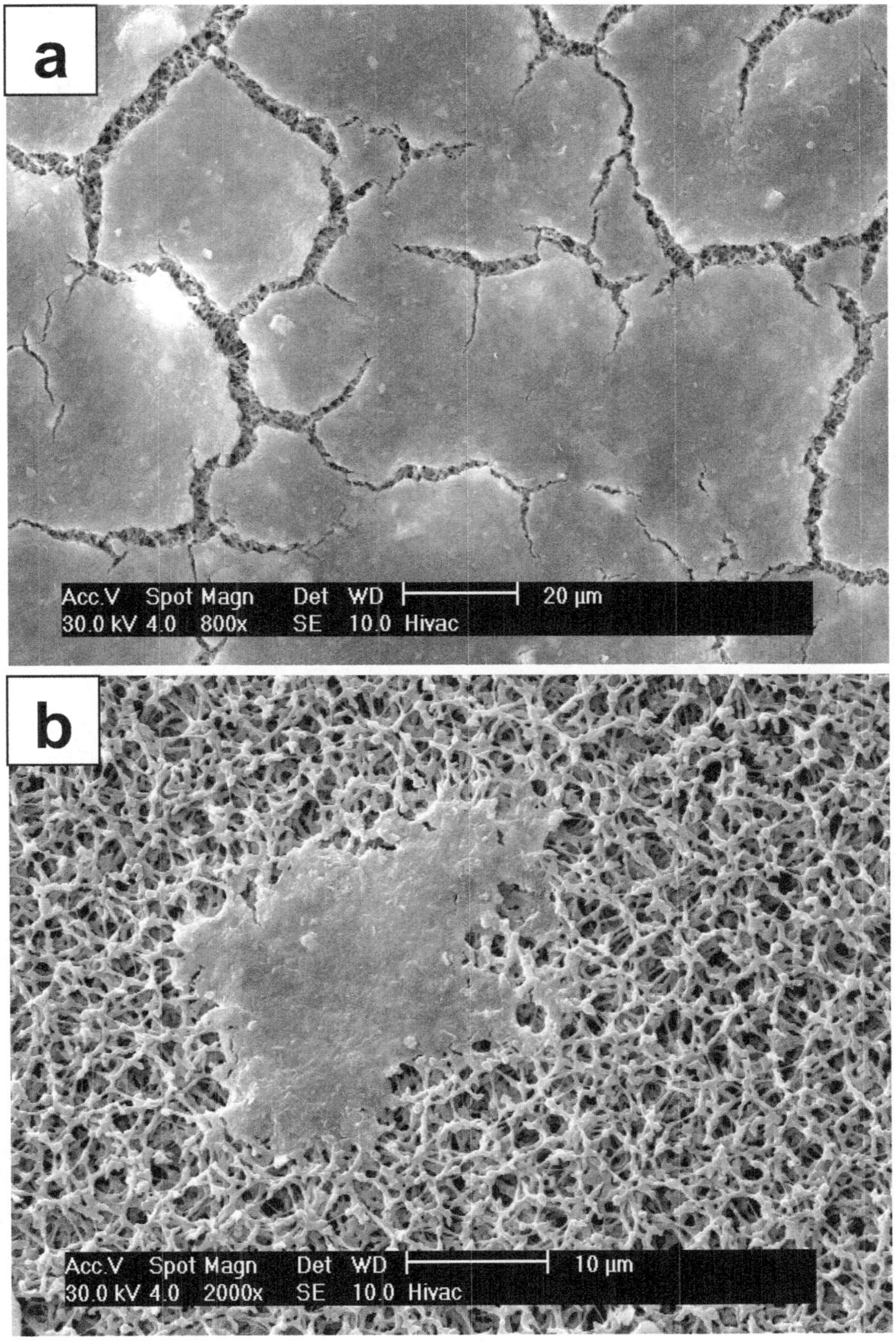

Figure 4.27 SEM images of material from the filter (a) influent and (b) effluent retained on a 0.45-µm filter pad at the East Water Purification Plant in Houston, TX during the unstressed operating period. The effluent sample was collected from Filter #9 during the middle of the filter run. The composition of the influent and effluent cakes is similarly 13% Al (Al_2O_3) and 8% Si (SiO) based on EDAX analysis, with the remaining being of an organic origin

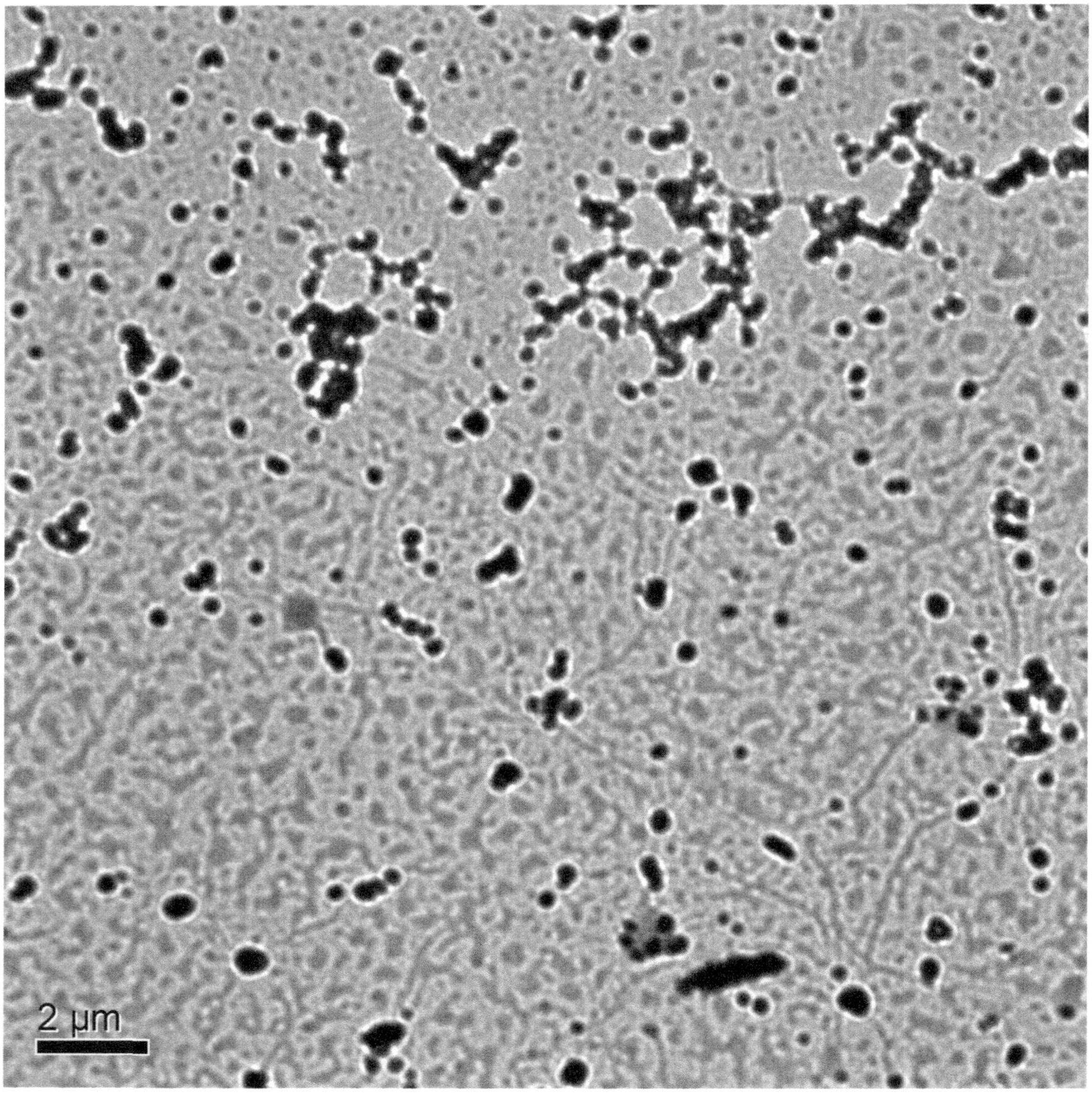

Figure 4.28 TEM image of ultrafine particles deposited on a copper mesh microgrid from the filter effluent during the stressed operating period at the East Water Purification Plant in Houston, TX. The effluent sample was collected from Filter #7 at the end of the filter run

Johnson County Water District, KS

Mass size distributions of organic and inorganic material present in Johnson County's filter influent and effluent are shown in Figure 4.29 – 4.32.

The organic material present in various size classes in the filter influent and effluent during the stressed sampling period is depicted in Figure 4.29. Organic material in the filter influent was present mostly in the ultrafine (39%) and dissolved + (54%) size fractions (Figure 4.29a). Approximately 7% is present in the settleable size class. In the filter effluent, all the organic material is present in the dissolved + (63% to 68%) and ultrafine (32% to 37%) fractions.

Inorganic mass fraction analysis of filter influent indicated that the material is evenly distributed between the dissolved + and ultrafine size fractions at approximately 47% in each class (Figure 4.30). Approximately 6% is present in the settleable size class. In the filter effluent, no material was observed to be in the settleable size class at any stage of filter run. Approximately 40% of the material was present in the dissolved + size range at the beginning of the filter run that increased to 48% at the middle of the run, before decreasing to 35% at the end of the run. The ultrafine fraction changed accordingly, as all the mass was present only in the dissolved + and ultrafine size classes. The material in the dissolved + and ultrafine size classes is similarly distributed between calcium, magnesium and silica for influent as well as effluent at various stages of filter run. The two hardness cations, calcium and magnesium, accounted for approximately 74% of the material, followed by silica at 27%. Aluminum was present at a concentration of 0.34 ppm while the concentration of iron in the filter influent was lower at 0.075 ppm. The aluminum principally existed in the filtrable size fraction (39%), followed by that in the ultrafine, settleable, and dissolved + size fractions, respectively. The iron on the other hand was only present mostly in the settleable size fraction (89%) with the remainder in the filtrable size fraction.

The mass size distribution of organic material present in the filter influent and effluent during the unstressed period is shown in Figure 4.31. Most of the organic material in the filter influent was present in the dissolved + (82%) and ultrafine (17%) fractions. In the filter effluent all the organic material was found to be present only in the dissolved + size range at all stages of filter run.

The mass size distribution of the inorganic material in the filter influent and effluent during the unstressed period is shown in Figure 4.32. In the filter influent and effluent, greater than 92% of the material is present in the dissolved + size range. At the end of filter run, approximately 7% of the material was evenly distributed between the ultrafine and settleable size fractions. The increase in ultrafine and settleable fractions correlated with increased calcium and magnesium concentrations in the effluent, suggesting formation or breakthrough of precipitates at this softening facility.

SEM images of the material in the filter influent and effluent that is retained on a 0.45-μm filter pad at the Johnson County water treatment facility are shown in Figure 4.33. The composition of the influent and effluent cakes was 22% Al (Al_2O_3), 46% Si (SiO), 22% Ca (CaO) and 4% Fe (Fe_2O_3) based on EDAX analysis. In addition, a TEM image of the filter effluent is given in Figure 4.34.

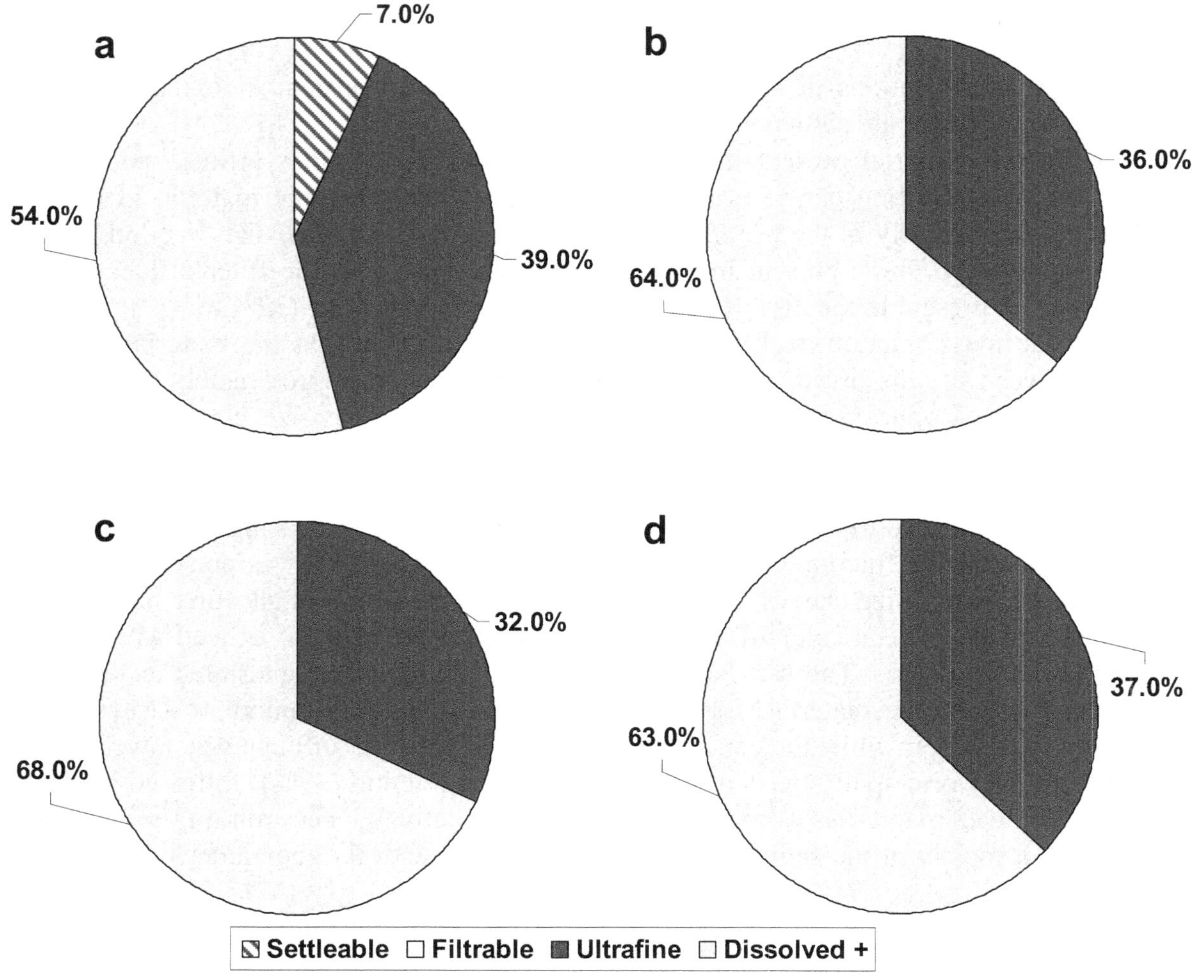

Figure 4.29 Size fractionation of the filter (a) influent and effluent TOC at the (b) start, (c) middle, and (d) end of the filter run for the Johnson County water treatment facility during the stressed sampling period (influent $[TOC]_T$ = 2.83 ppm, effluent $[TOC]_T$ = 2.63 ppm)

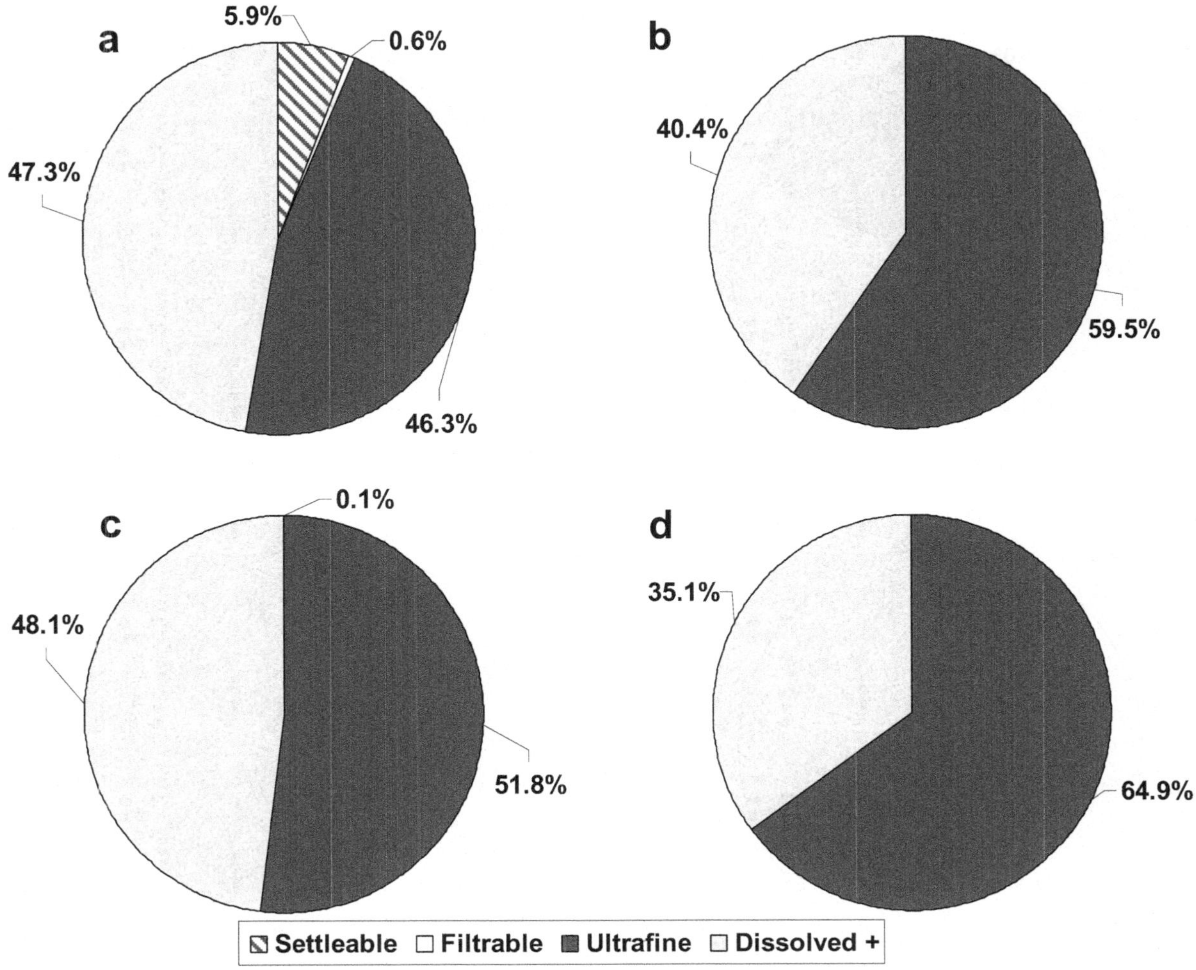

Figure 4.30 Size fractionation of the filter (a) influent and effluent elemental composition at the (b) start, (c) middle, and (d) end of the filter run for the Johnson County water treatment facility during the stressed sampling period (influent $[Metals]_T$ = 58.61 ppm, effluent $[Metals]_T$ = 52.22 ppm)

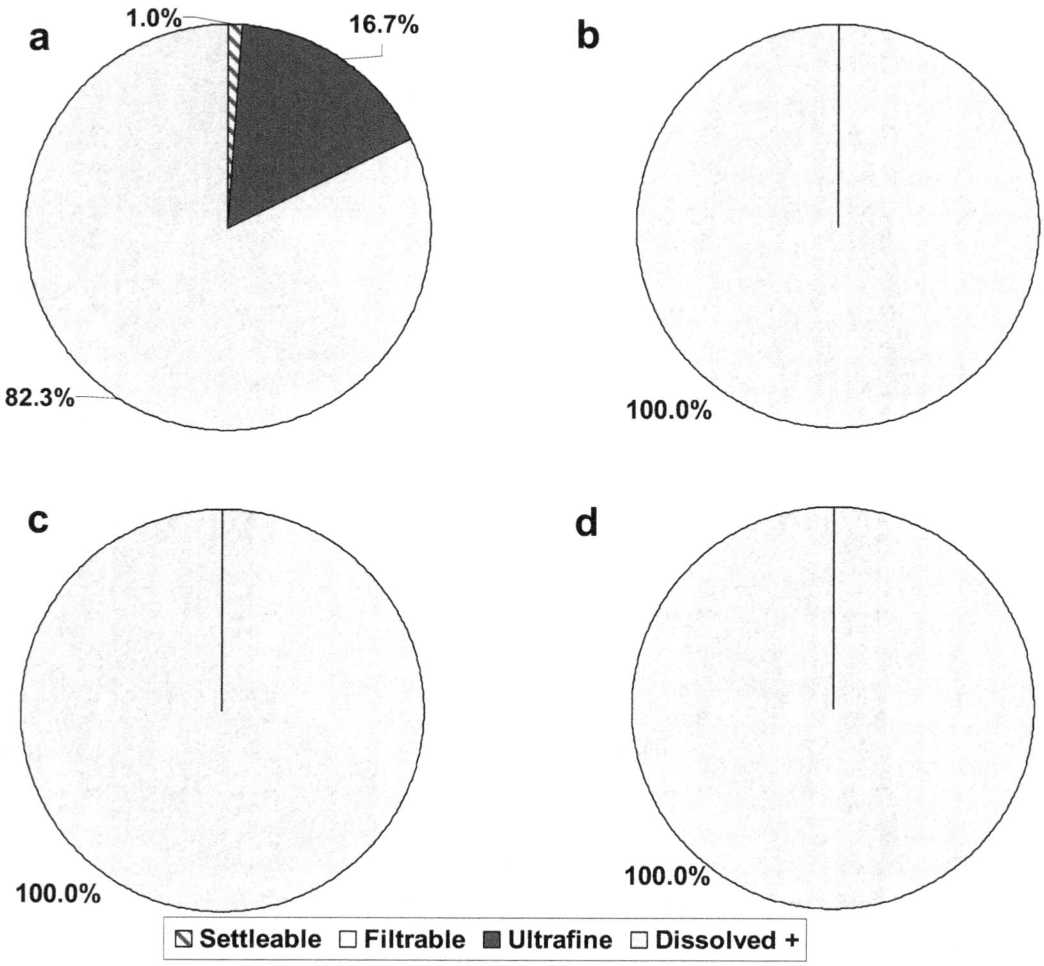

Figure 4.31 Size fractionation of the filter (a) influent and effluent TOC at the (b) start, (c) middle, and (d) end of the filter run for the Johnson County water treatment facility during the unstressed sampling period (influent $[TOC]_T$ = 2.22 ppm, effluent $[TOC]_T$ = 2.07 ppm)

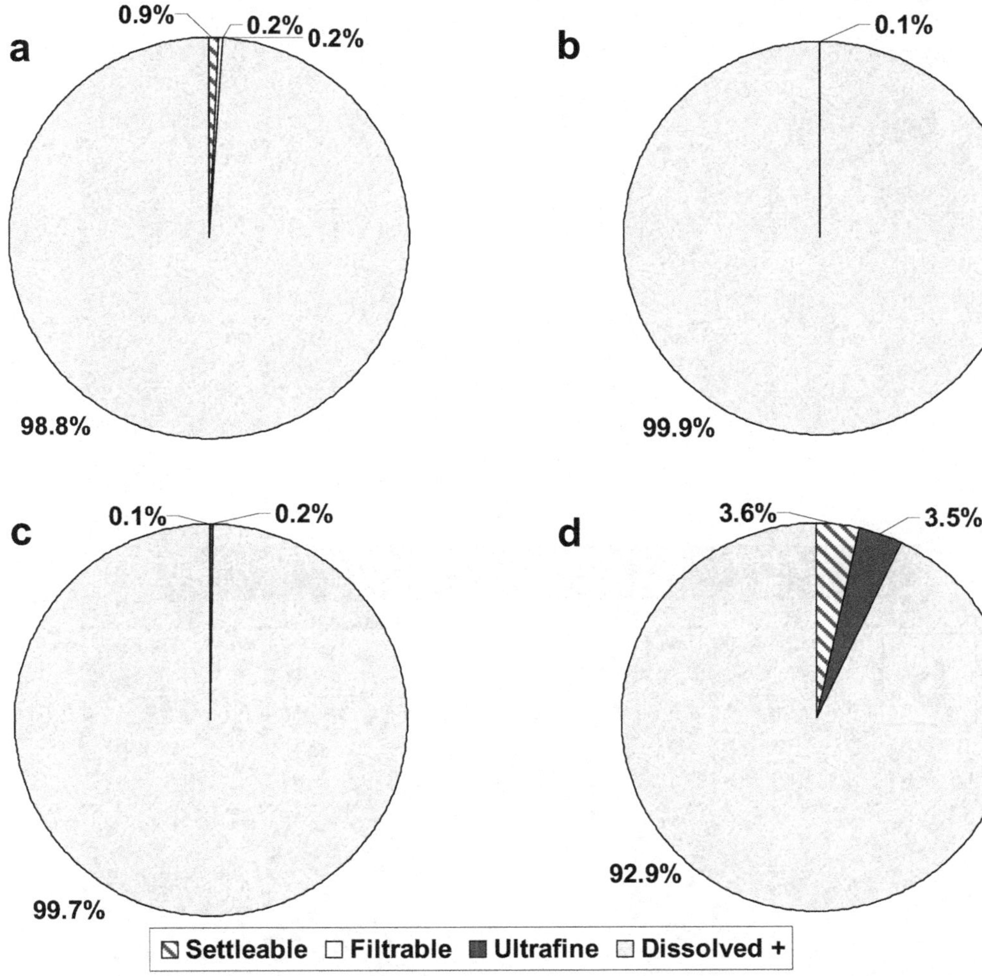

Figure 4.32 Size fractionation of the filter (a) influent and effluent elemental composition at the (b) start, (c) middle, and (d) end of the filter run for the Johnson County water treatment facility during the unstressed sampling period (influent $[Metals]_T$ = 43.33 ppm, effluent $[Metals]_T$ = 45.76 ppm)

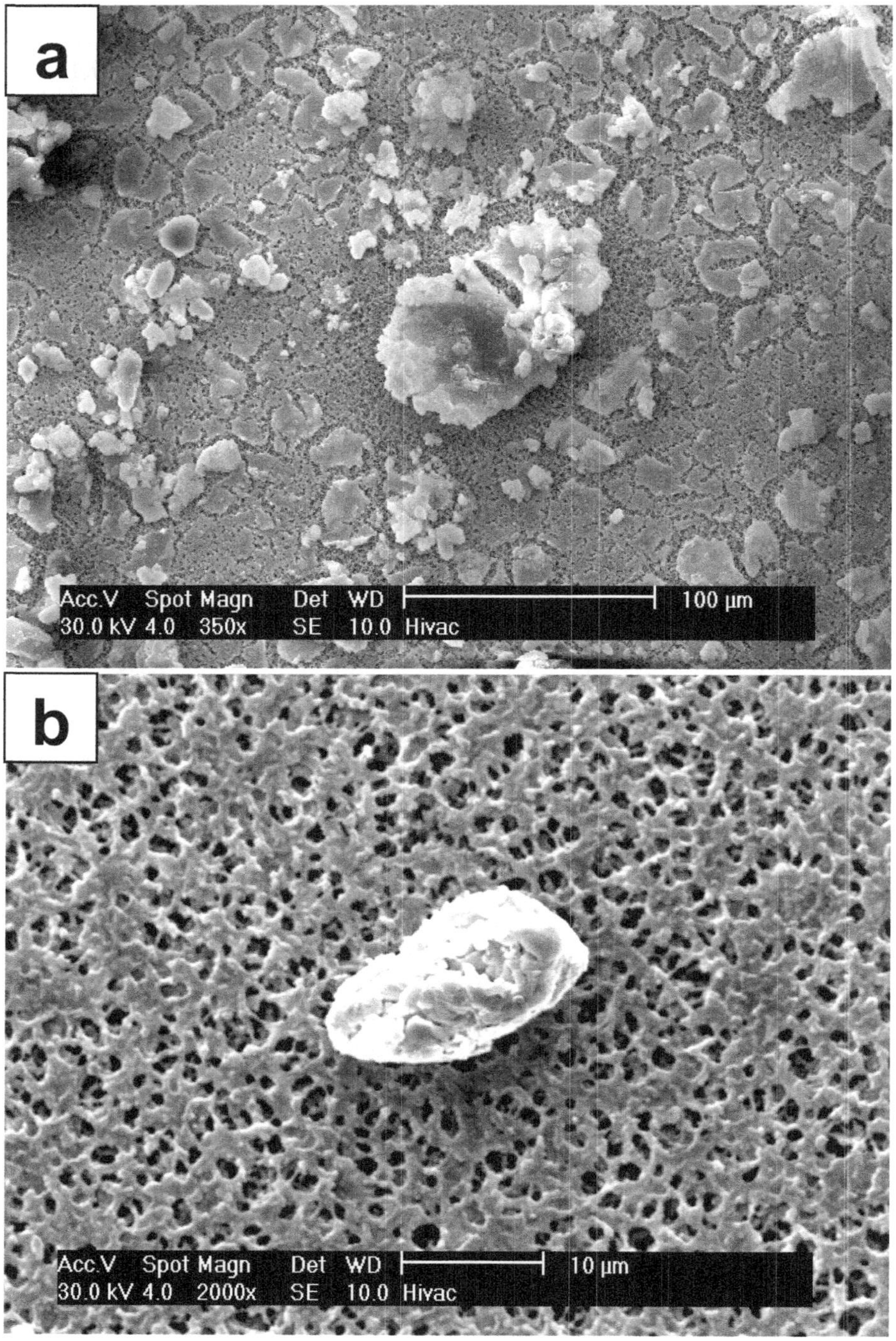

Figure 4.33 SEM images of material from the filter (a) influent and (b) effluent retained on a 0.45-µm filter pad at the Johnson County water treatment plant during the stressed operating period. The effluent sample was collected from Filter #13 during the middle of the filter run. The composition of the influent and effluent cakes is 22% Al (Al_2O_3), 46% Si (SiO), 22% Ca (CaO) and 4% Fe (Fe_2O_3) based on EDAX analysis. The particle in (b) was determined to be an organic composite

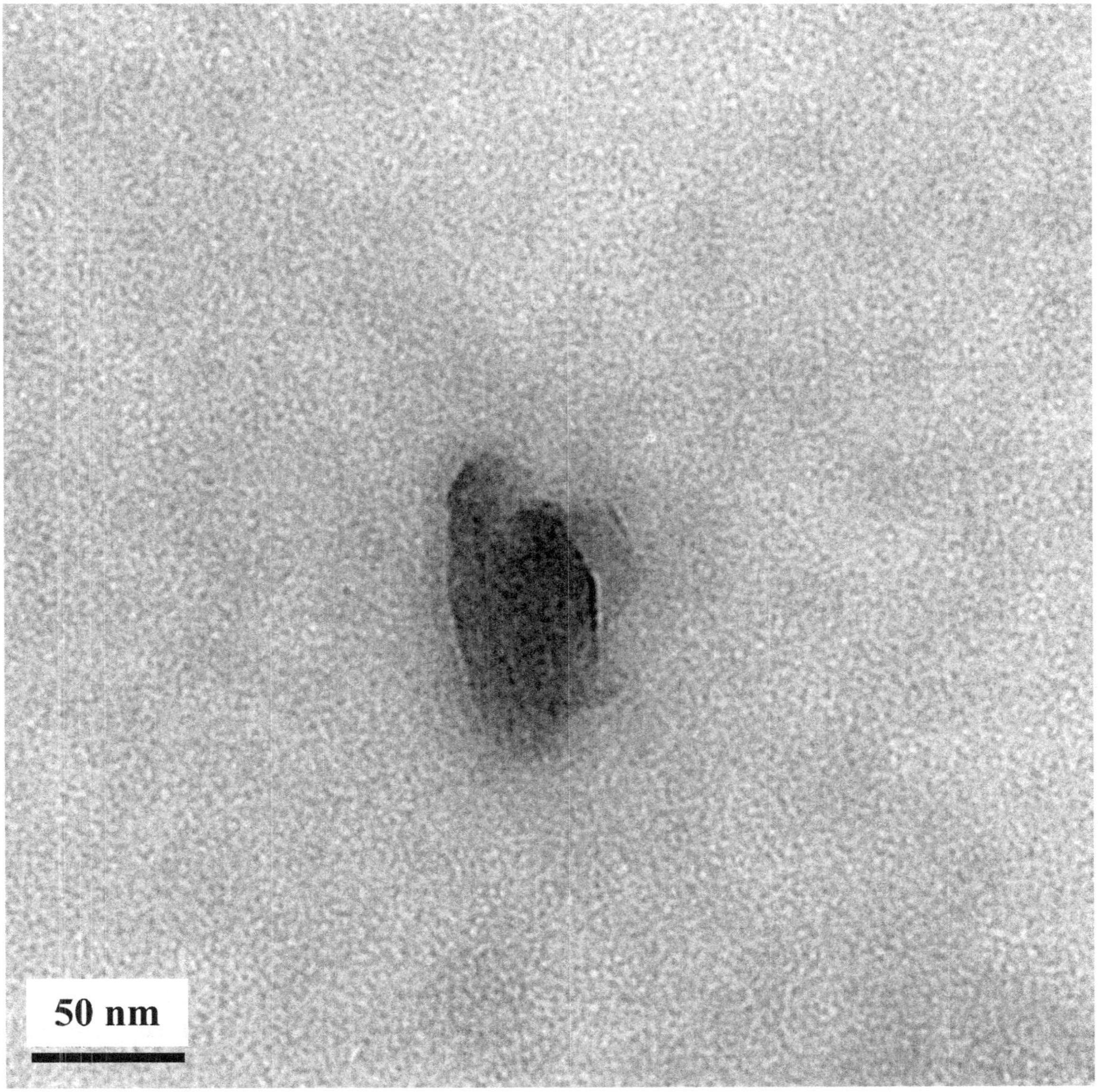

Figure 4.34 TEM image of ultrafine particles deposited on a copper mesh microgrid from the filter effluent during the unstressed operating period at the Johnson County water treatment facility. The effluent sample was collected at Filter #18 during the middle of the filter run

Modesto Irrigation District, CA

The organic mass size fractions for the filter influent and effluent during the stressed filter period are shown in Figure 4.35 for the Modesto Water Treatment Plant. Most of the organic material in the filter influent and effluent was present in the dissolved + size range (74% to 80%). The ultrafine fraction was 8% in the filter influent and varied over a small range (12 to 14%) in the filter effluent during the course of filter run. The filtrable fraction was highest in the filter influent (13%), while the material in this size class in the filter effluent was 2% throughout the filter run. The settleable fraction in the influent and effluent ranged from 4 to 7%.

The distribution of inorganic material into relative size fractions for the filter influent and effluent under stressed conditions are shown in Figure 4.36. Inorganic material was primarily present in the dissolved + (68%) and ultrafine (27%) size fractions in the filter influent. This same trend was observed for the filter effluent during the three stages of the filter run. However, the relative amount of material in the dissolved + size range increased and that in the ultrafine size range decreased. The composition of inorganic material in both the influent and effluent is similar, with significant amounts of silica, calcium and magnesium. Small amounts of iron, aluminum and manganese were also detected.

The mass distribution of organic matter in the filter influent and effluent during the unstressed period is shown in Figure 4.37. The organic material in the influent was distributed between dissolved + (46%), settleable (31%), ultrafine (19%) and filtrable (5%). However, in the filter effluent, most of the organic matter was present in the dissolved + size range (approximately 70%). The settleable fraction was lower compared to the influent (17 to 22%). The filtrable fraction was higher at the beginning of the run at 11% and decreased to approximately 5% during the later stages of filter run.

The inorganic material mass distribution in the filter influent and effluent is shown in Figure 4.38. In the filter influent, the inorganic material was primarily present in the dissolved + (78%) and ultrafine (19%) size classes. In the filter effluent, approximately 86% of the inorganic material was present in the dissolved + size class throughout the filter run. The ultrafine and settleable fractions were also relatively unchanged at approximately 7% and 5% respectively.

SEM images of the materials in the filter influent and effluent retained on a 0.45-μm filter pad at the Modesto water treatment facility are shown in Figure 4.39. The composition the influent cake is 12% Al (Al_2O_3) and 21% Si (SiO) with the remainder being of an organic origin based on EDAX analysis. In addition, a TEM image of the filter effluent is given in Figure 4.40. It should be noted that particles such as these are normally present in the water and do not necessarily pose any health effects. These pictures are included to indicate the morphology of the particles that are normally present in drinking water.

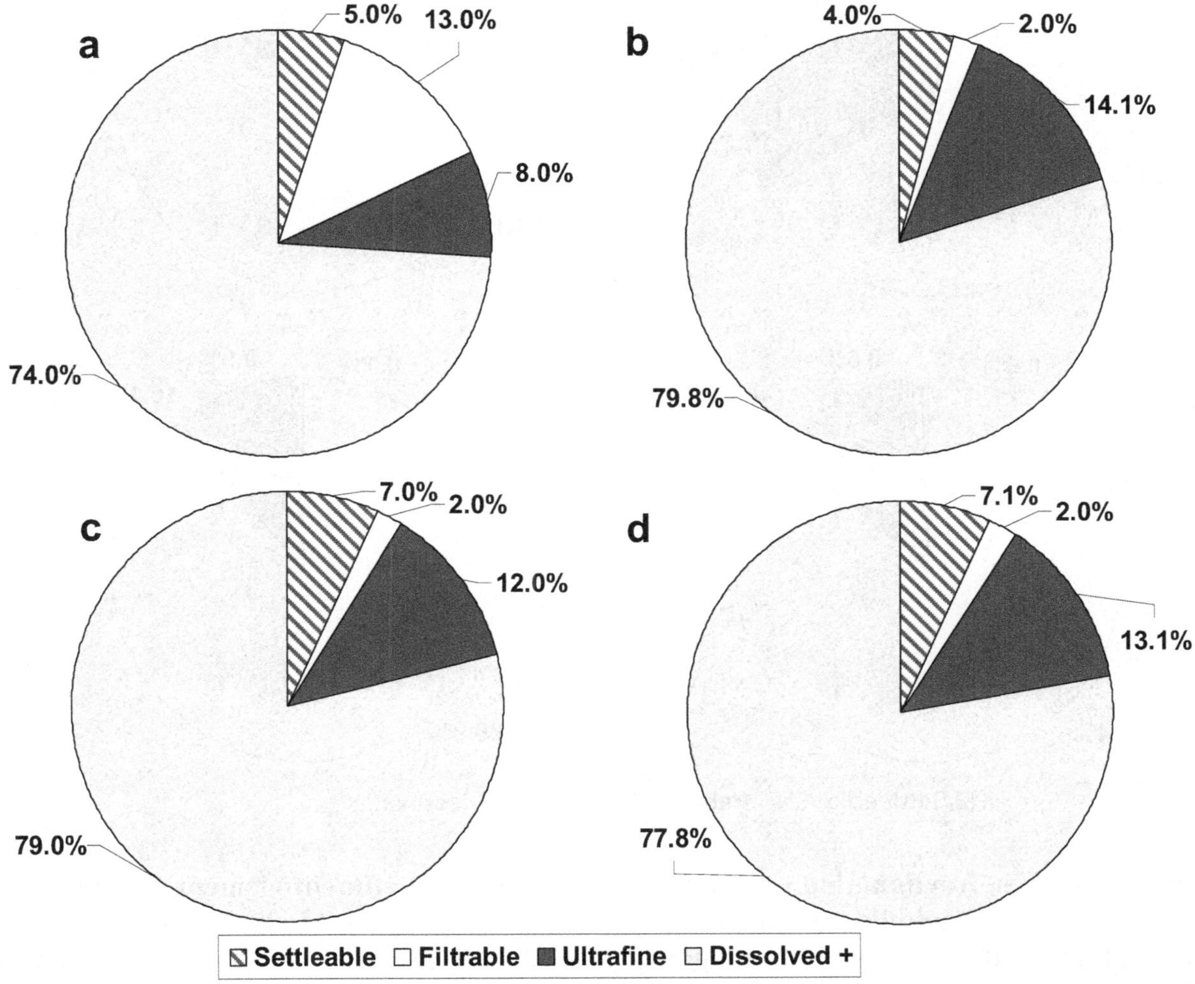

Figure 4.35 Size fractionation of the filter (a) influent and effluent TOC at the (b) start, (c) middle, and (d) end of the filter run for the Modesto water treatment facility during the stressed sampling period (influent $[TOC]_T$ = 1.47 ppm, effluent $[TOC]_T$ = 1.39 ppm). The concentration of TOC in settleable, filtrable, ultrafine and dissolved + size classes, for each sample are, respectively, as follows: (a) 0.07 mg/L, 0.19 mg/L, 0.11 mg/L, 1.08 mg/L; (b) 0.06 mg/L, 0.03 mg/L, 0.20 mg/L, 1.11 mg/L; (c) 0.1 mg/L, 0.03 mg/L; 0.18 mg/L, 1.08 mg/L; (d) 0.1 mg/L; 0.03 mg/L, 0.17 mg/L, 1.10 mg/L.

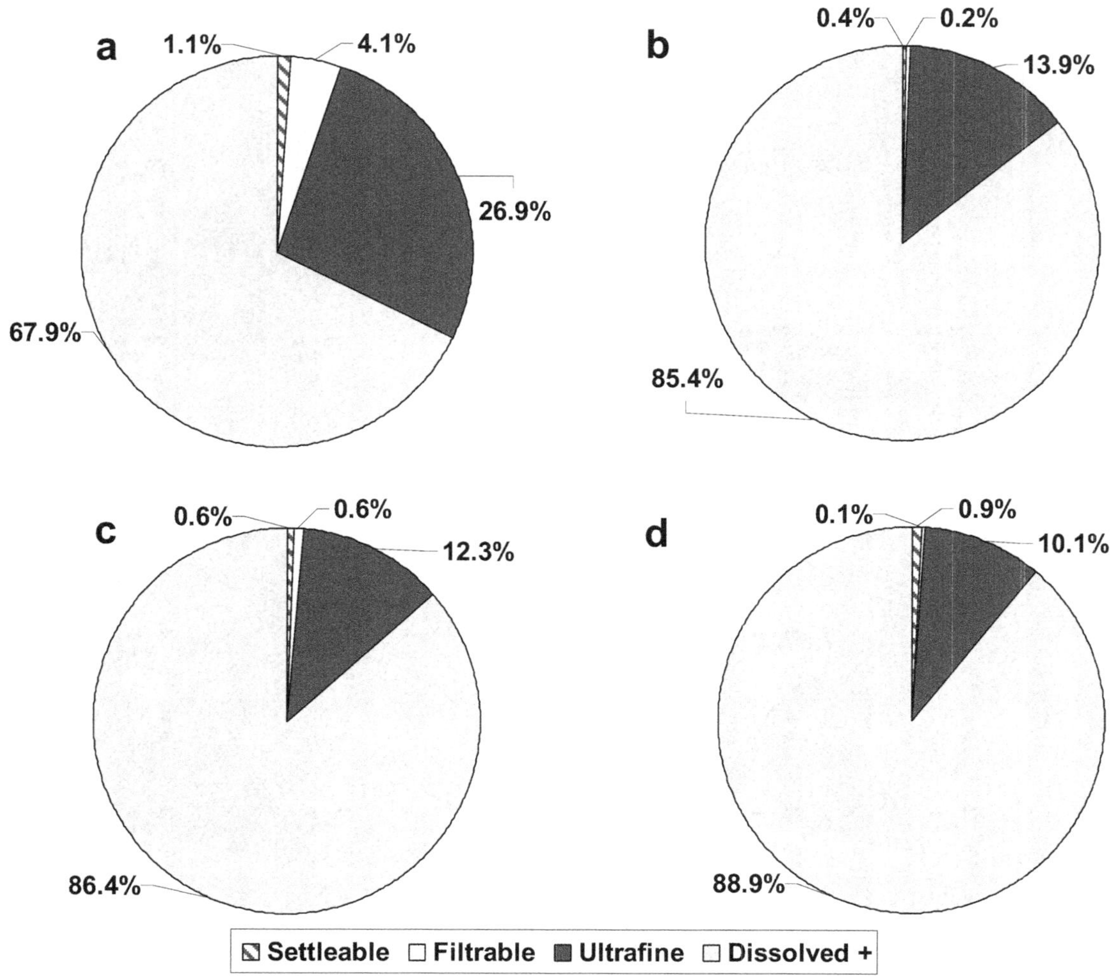

Figure 4.36 Size fractionation of the filter (a) influent and effluent elemental composition at the (b) start, (c) middle, and (d) end of the filter run for the Modesto water treatment facility during the stressed sampling period (influent $[Metals]_T$ = 12.39 ppm, effluent $[Metals]_T$ = 11.93 ppm); The concentration of inorganic material in settleable, filtrable, ultrafine and dissolved + size classes, for each sample are, respectively, as follows: (a) 0.14 mg/L, 0.51 mg/L, 3.33 mg/L, 8.41 mg/L; (b) 0.05 mg/L, 0.02 mg/L, 1.66 mg/L, 10.19 mg/L; (c) 0.07 mg/L, 0.07 mg/L; 1.47 mg/L, 10.31 mg/L; (d) 0.01 mg/L; 0.11 mg/L, 1.2 mg/L, 10.61 mg/L.

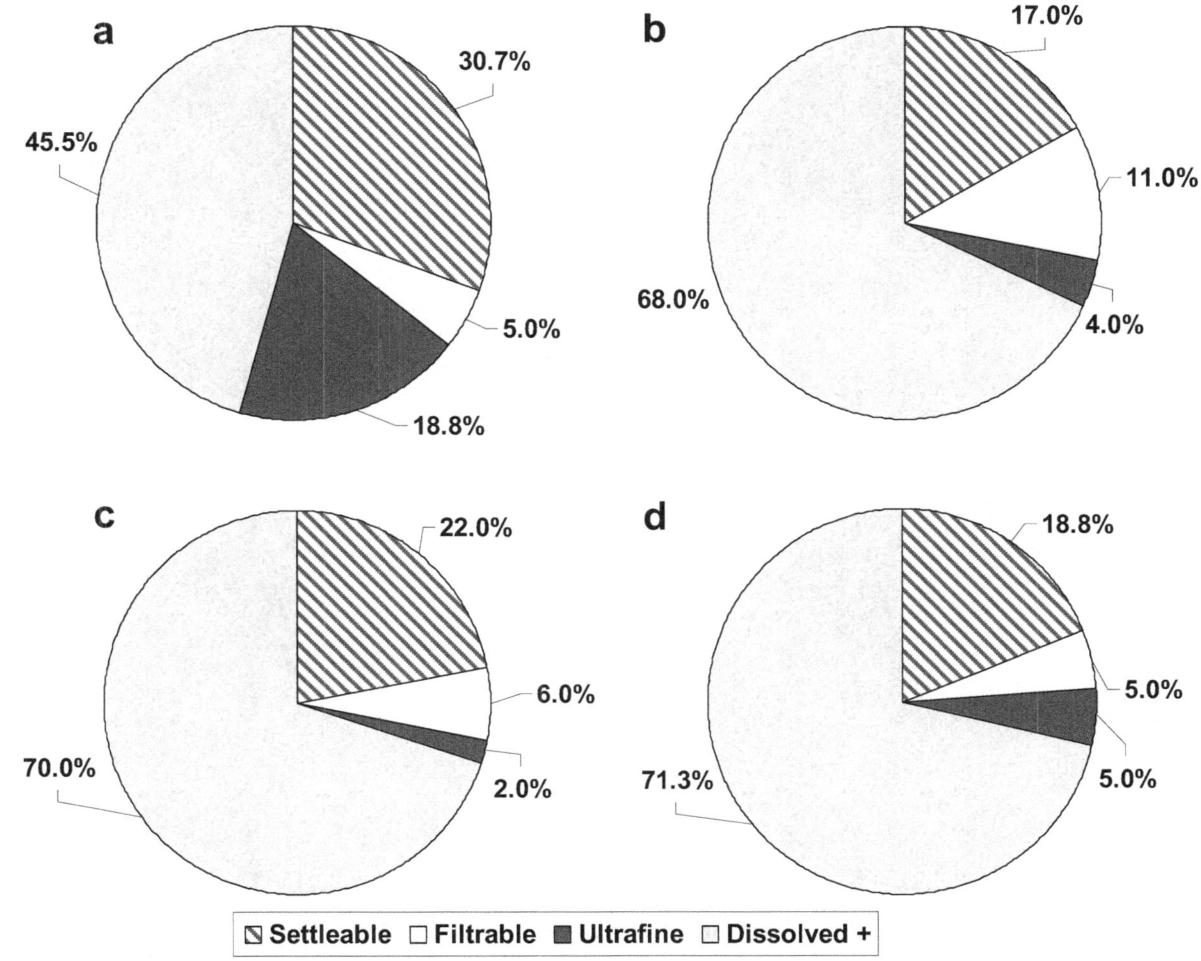

Figure 4.37 Size fractionation of the filter (a) influent and effluent TOC at the (b) start, (c) middle, and (d) end of the filter run for the Modesto water treatment facility during the unstressed sampling period (influent $[TOC]_T$ = 1.64 ppm, effluent $[TOC]_T$ = 1.48 ppm); The concentration of TOC in settleable, filtrable, ultrafine and dissolved + size classes, for each sample are, respectively, as follows: (a) 0.52 mg/L, 0.08 mg/L, 0.31 mg/L, 0.74 mg/L; (b) 0.25 mg/L, 0.16 mg/L, 0.06 mg/L, 1.01 mg/L; (c) 0.33 mg/L, 0.09 mg/L; 0.03 mg/L, 1.04mg/L; (d) 0.28 mg/L; 0.07 mg/L, 0.07 mg/L, 1.06 mg/L.

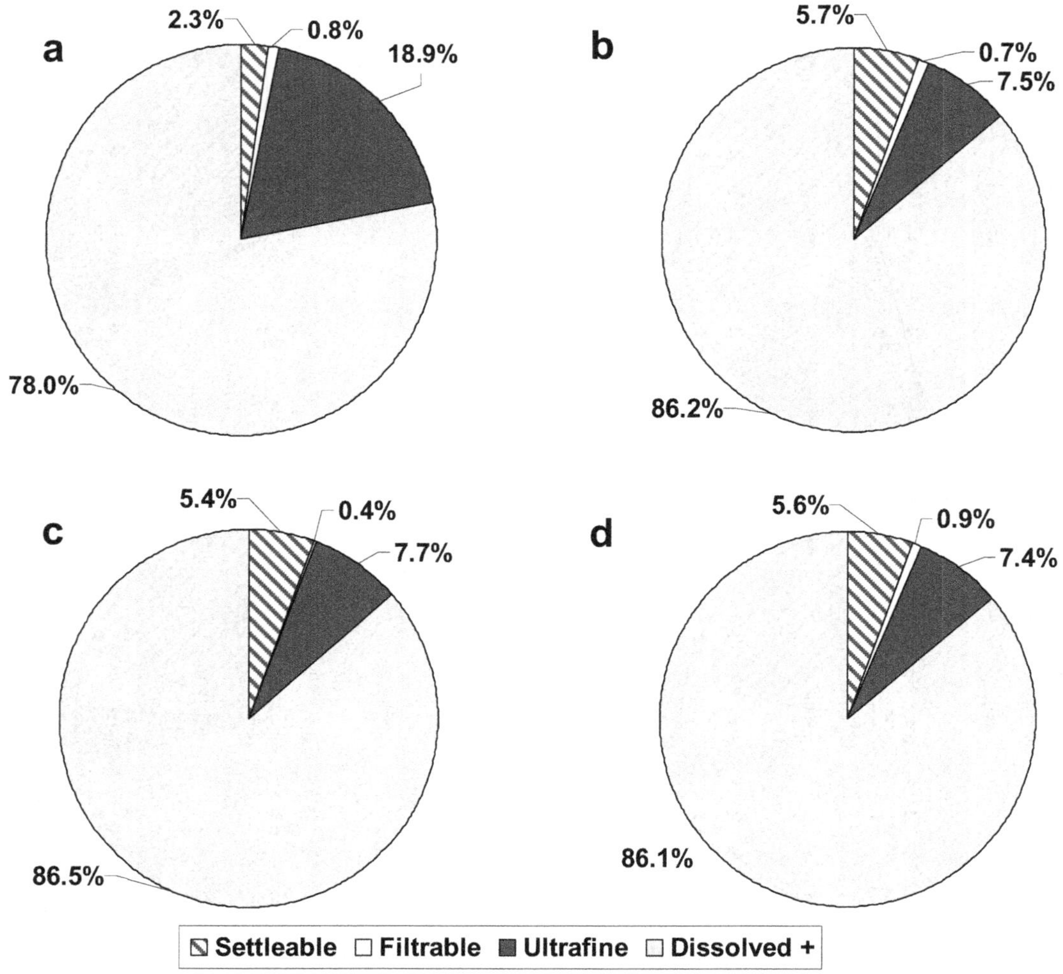

Figure 4.38 Size fractionation of the filter (a) influent and effluent elemental composition at the (b) start, (c) middle, and (d) end of the filter run for the Modesto water treatment facility during the unstressed sampling period (influent $[Metals]_T$ = 11.96 ppm, effluent $[Metals]_T$ = 10.83 ppm); The concentration of inorganic material in settleable, filtrable, ultrafine and dissolved + size classes, for each sample are, respectively, as follows: (a) 0.28 mg/L, 0.10 mg/L, 2.26 mg/L, 9.33 mg/L; (b) 0.62 mg/L, 0.06 mg/L, 0.81 mg/L, 9.33 mg/L; (c) 0.58 mg/L, 0.04 mg/L; 0.83 mg/L, 9.37 mg/L; (d) 0.61 mg/L; 0.1 mg/L, 0.8 mg/L, 9.32 mg/L.

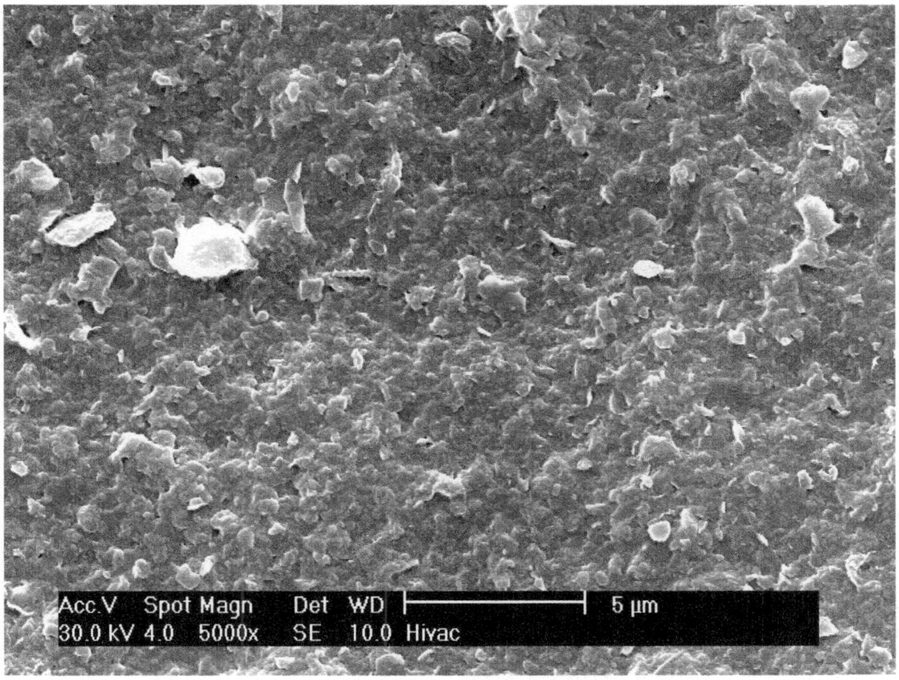

Figure 4.39 SEM images of material from the filter influentretained on a 0.45-µm filter pad at the Modesto Irrigation District water treatment plant during the stressed operating period. The composition of the influent cake is 12% Al (Al_2O_3) and 21% Si (SiO) with the remainder being of an organic origin based on EDAX analysis.

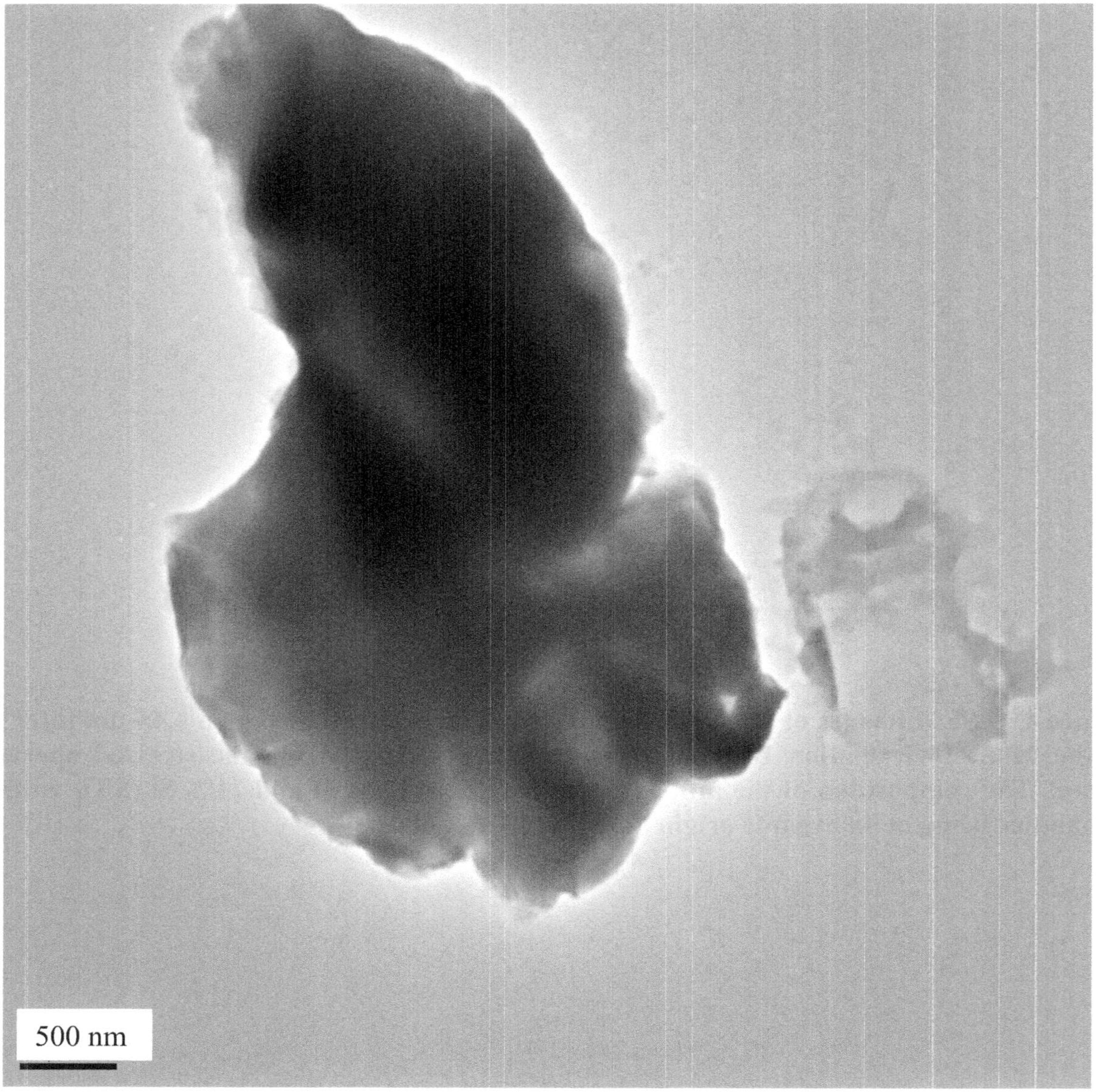

Figure 4.40 TEM image of an ultrafine particle deposited on a copper mesh microgrid from the filter effluent during the stressed operating period at the Modesto Irrigation District water treatment facility. The sample was collected from Filter #4

Southern Nevada Water Authority

Size fractionations of organic and inorganic materials in the Southern Nevada Water Authority's filter influent and effluent are given in Figure 4.41 – 4.44 during the stressed and unstressed periods.

The size distributions of organic material present in the filter influent and effluent during the unstressed period are shown in Figure 4.41. The organic material in the influent was present in only dissolved (81%) and ultrafine (19%) size classes. At the beginning of the filter run, most of the organic material was present in the dissolved + size range (67%), followed by settleable (20%,) and filtrable (13%). At the middle of the run, the dissolved + fraction increased to 83%, and the settleable fraction decreased to 12%. None of the organic material was present in the filtrable fraction, and approximately 5% of the material was determined to be in ultrafine fraction. Towards the end of the filter run, 88% of the organic material was in the dissolved + size class, 11% in filtrable class and 1% in settleable.

The inorganic material size fractionations in the filter influent and effluent during the unstressed period are shown in Figure 4.42. Most the material in the influent and effluent was present in the dissolved + size range (approximately 97%). Similar trend was observed during stressed period, where greater than 97% of the inorganic material was found to be in dissolved + size range (Figure 4.44).

The organic material size distribution in the filter influent and effluent during stressed period is shown in Figure 4.43. In the influent, most of the organic material was present in the dissolved + size range (80%) followed by 12% in the settleable size range, 5% in the filtrable range and 3% in the ultrafine range. At the beginning of the filter run, 7% of the inorganic material was in the ultrafine range, with the rest in the dissolved + size class. During later stages of filter run, all the material was found to be in the dissolved + size range.

SEM images of the material in the influent and effluent that is retained on a 0.45-μm filter pad at the Southern Nevada water treatment facility is shown in Figure 4.45.

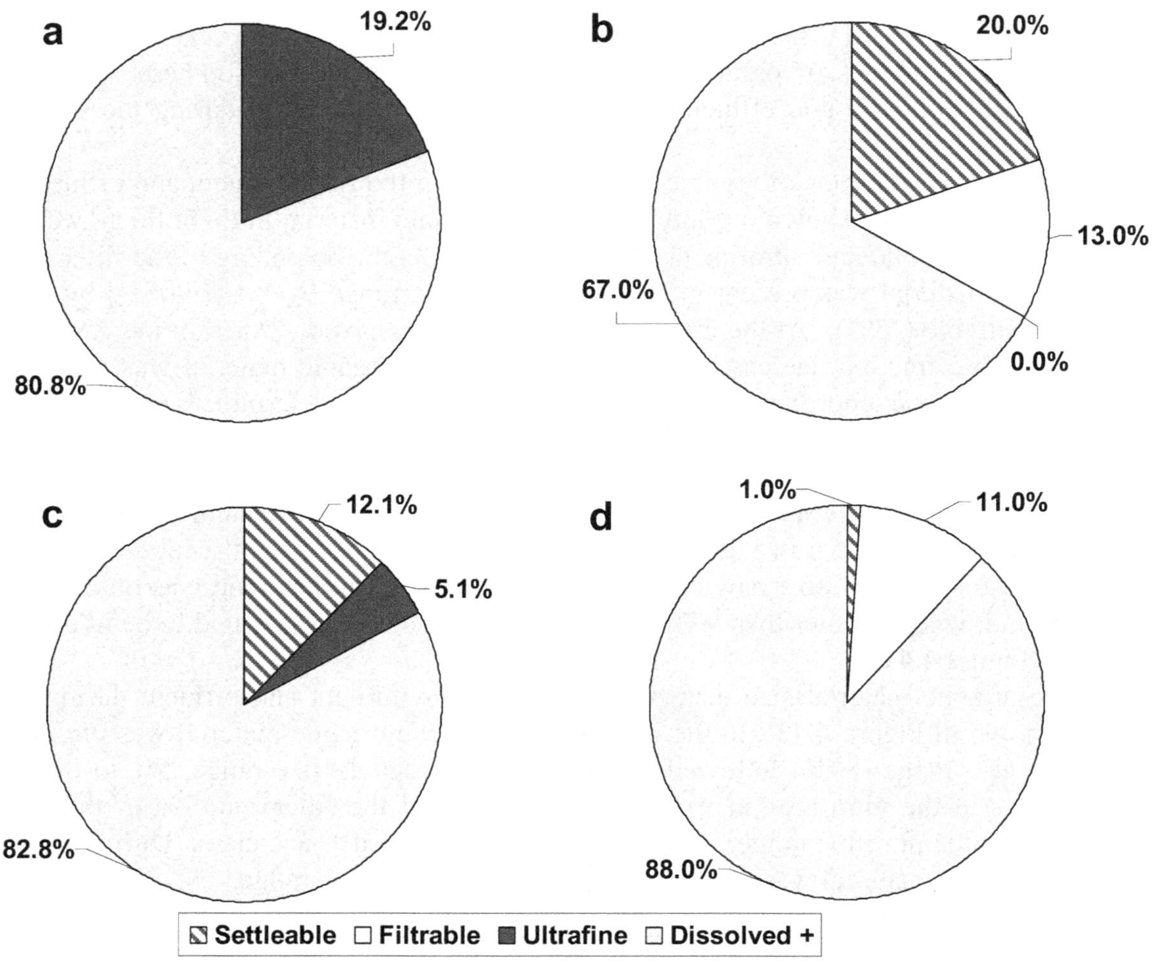

Figure 4.41 Size fractionation of the filter (a) influent and effluent TOC at the (b) start, (c) middle, and (d) end of the filter run for the Southern Nevada water treatment facility during the unstressed sampling period (influent $[TOC]_T = 3.08$ ppm, effluent $[TOC]_T = 3.69$ ppm)

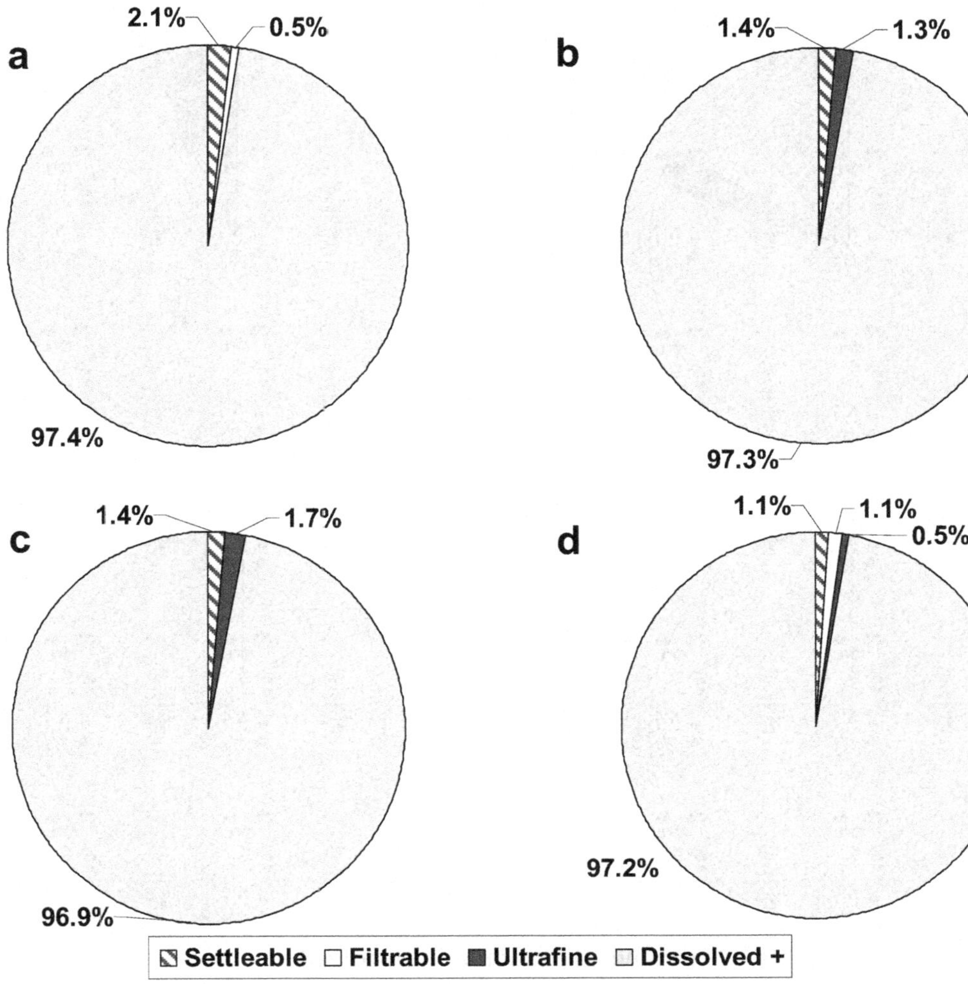

Figure 4.42 Size fractionation of the filter (a) influent and effluent elemental composition at the (b) start, (c) middle, and (d) end of the filter run for the Southern Nevada water treatment facility during the unstressed sampling period (influent $[Metals]_T$ = 116.76 ppm, effluent $[Metals]_T$ = 115.48 ppm)

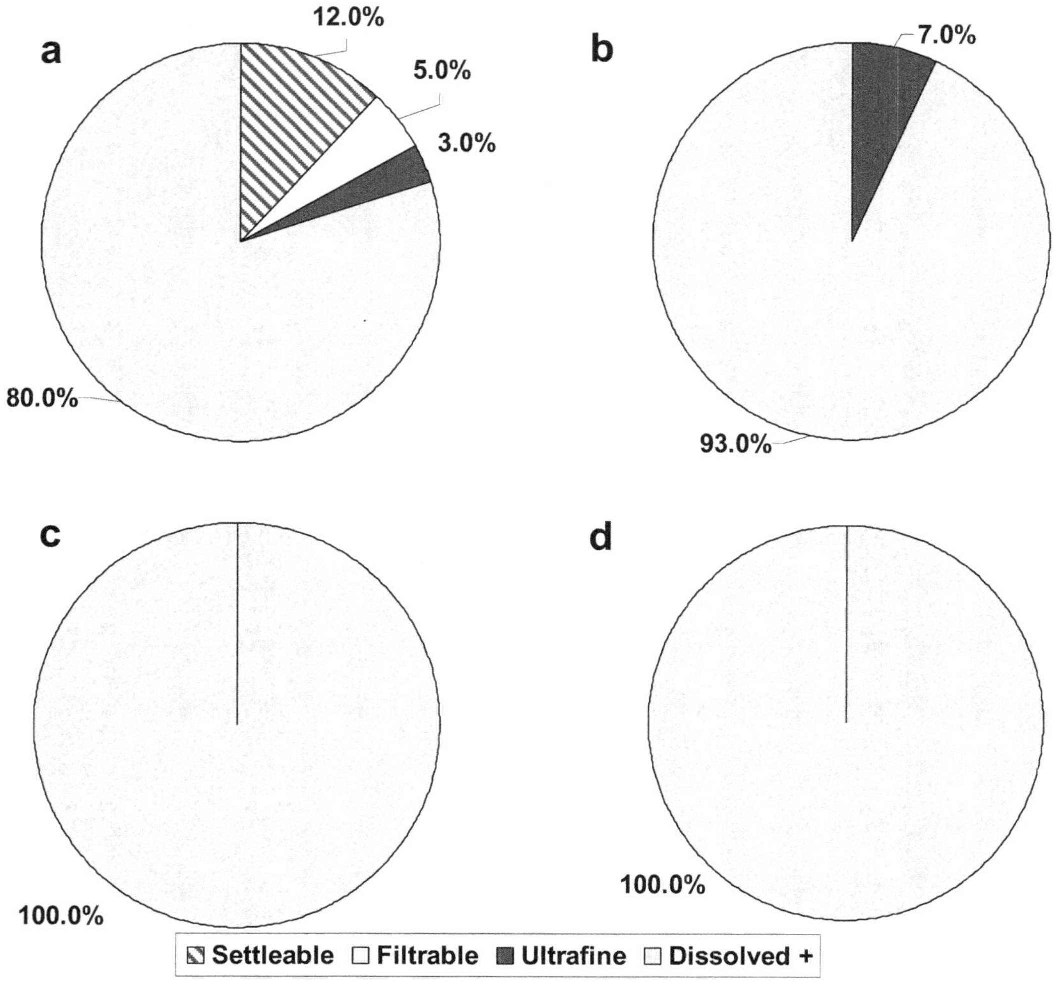

Figure 4.43 Size fractionation of the filter (a) influent and effluent TOC at the (b) start, (c) middle, and (d) end of the filter run for the Southern Nevada water treatment facility during the stressed sampling period (influent $[TOC]_T$ = 3.16 ppm, effluent $[TOC]_T$ = 2.96 ppm)

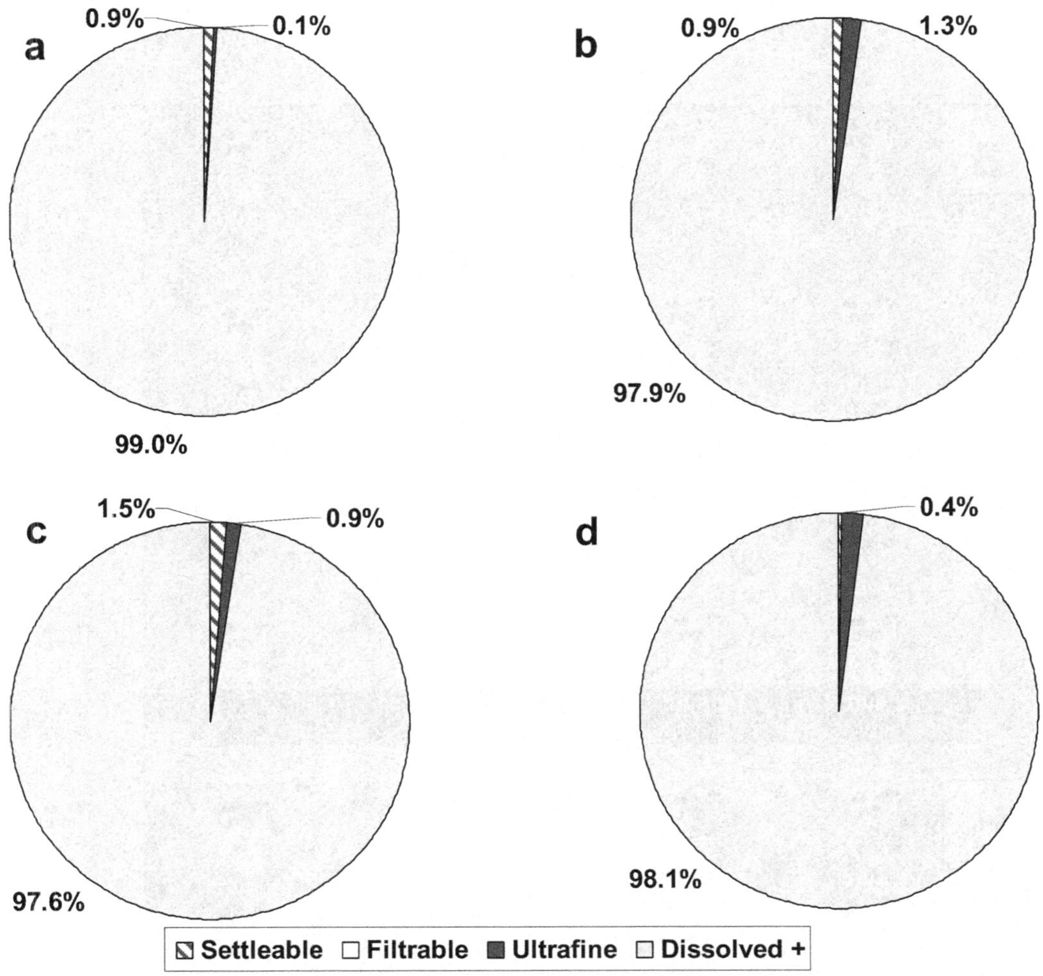

Figure 4.44 Size fractionation of the filter (a) influent and effluent elemental composition at the (b) start, (c) middle, and (d) end of the filter run for the Southern Nevada water treatment facility during the stressed sampling period (influent $[Metals]_T$ = 116.08 ppm, effluent $[Metals]_T$ = 115.48 ppm)

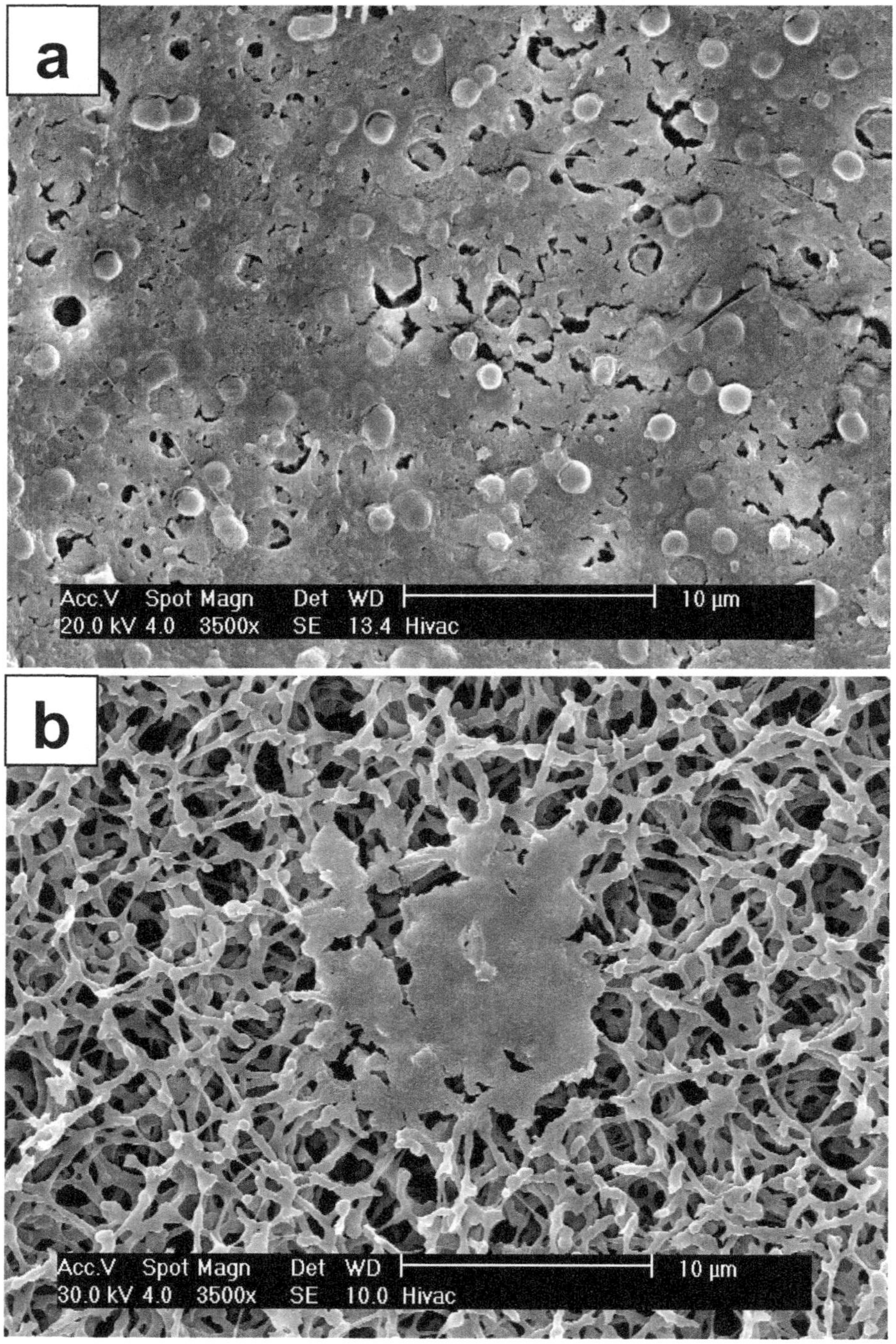

Filter 4.45 SEM images of material from the filter (a) influent and (b) effluent retained on a 0.45-µm filter pad at the Southern Nevada water treatment plant during the unstressed operating period. The effluent sample was collected from Filter #8 during the middle of the filter run. The composition of the (a) influent cake is 20% Ca (CaO) with the remainder being of an organic origin

Salem Public Works Department, OR

The mass distribution of organic and inorganic material in the influent and effluent of City of Salem's filters are shown in Figures 4.46 and 4.47.

The distribution of organic material in various size classes for the filter influent and effluent are shown in Figure 4.46. Unlike most other facilities sampled, the organic material in both the filter influent and effluent appears to be more evenly distributed in all the size classes. The dissolved + and filtrable size fractions are lower in the effluent compared to the influent, while the settleable fraction increased significantly (from 11% to 26%) in the effluent. The ultrafine fraction increased slightly from approximately 25% to 28%.

The distribution of the inorganic material in the filter influent and effluent are shown in Figure 4.47. Bulk of the inorganic material is present in the dissolved + size fraction. Silica (58%), calcium (33%) and magnesium (8%) contributed to most of the inorganic material in the influent. In the effluent, the composition was similar with the same elements dominating the inorganic matter (silica (54%), calcium (37%) and magnesium (9%).

SEM images of influent and effluent retained on a 0.45-μm filter pad at the of Salem water treatment facility are shown in Figure 4.48 and 4.49, and a TEM image of the Salem water treatment facility filter effluent is given in Figure 4.50.

Significant numbers of large filter-clogging diatoms (up to 100 μm in size) were identified under the SEM, as *Cyclotella*, and *Melosira* (Figure 4.48) (Aquatic organism identification referred to Algae color plates in *Standards Methods For the Examination of Water and Wastewater,* 14th Edition, 1975. APHA – AWWA- WPCF). Nanometer sized particles were also recognized using TEM imaging (Figure 4.50).

A turbidimeter was unable to identify these diatoms, due to the transparent nature of the cell walls of these microorganisms (Turbidity reported for Santiam River Water is 2.4 NTU). It is likely that most of these diatoms originated from sediment accumulated in the raw water intake pipe.

No particle deposit was observed on the filter-pad through which filter effluent was filtered (Figure 4.49). This observation agrees with the very low value of effluent turbidity (0.08 NTU).

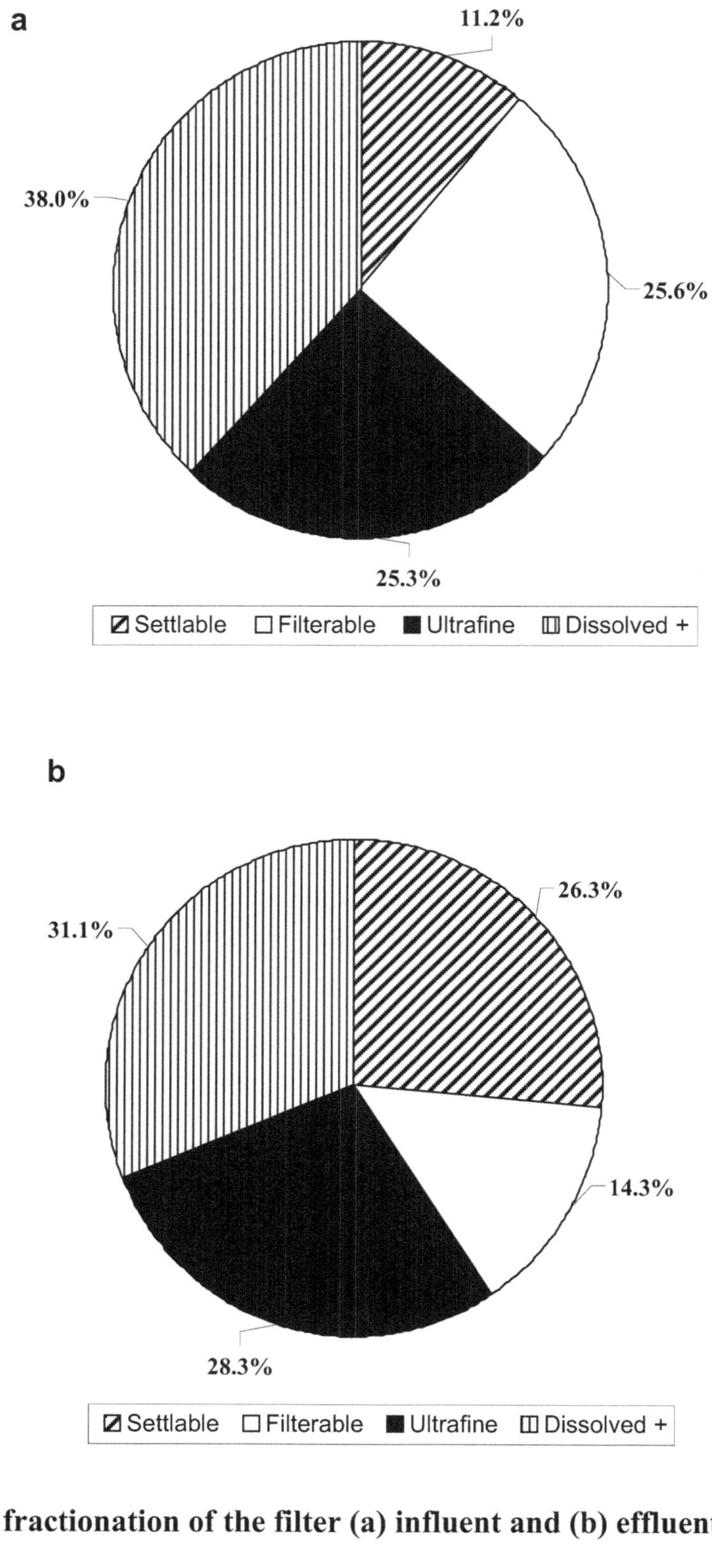

Figure 4.46 Size fractionation of the filter (a) influent and (b) effluent TOC of Salem's filter (influent $[TOC]_T = 0.67$ ppm, effluent $[TOC]_T = 0.66$ ppm).

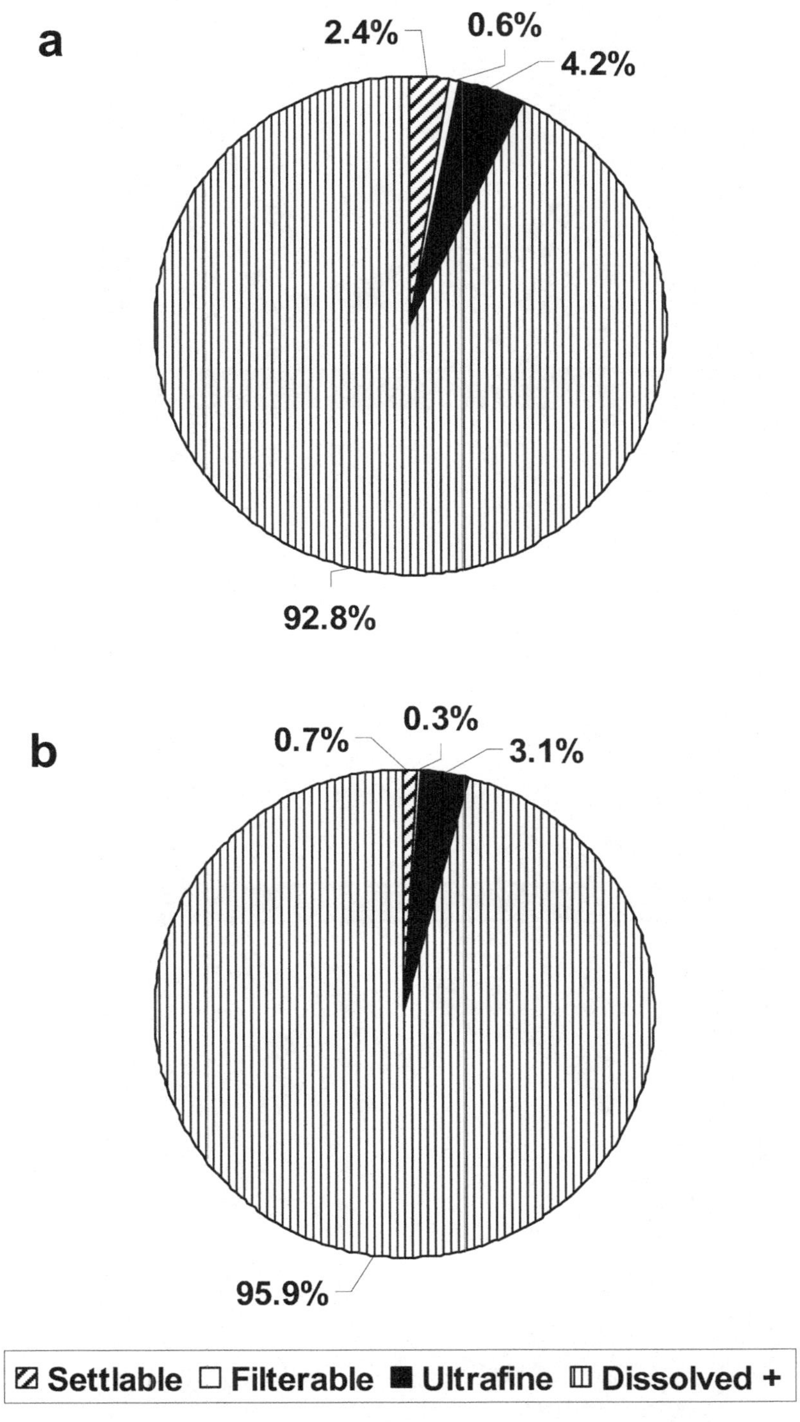

Figure 4.47 Relative size distributions of inorganics in the filter (a) influent and (b) effluent at the Salem water treatment plant during the stressed operating period (influent $[Metals]_T$ = 15.71 ppm, effluent $[Metals]_T$ = 15.50 ppm)

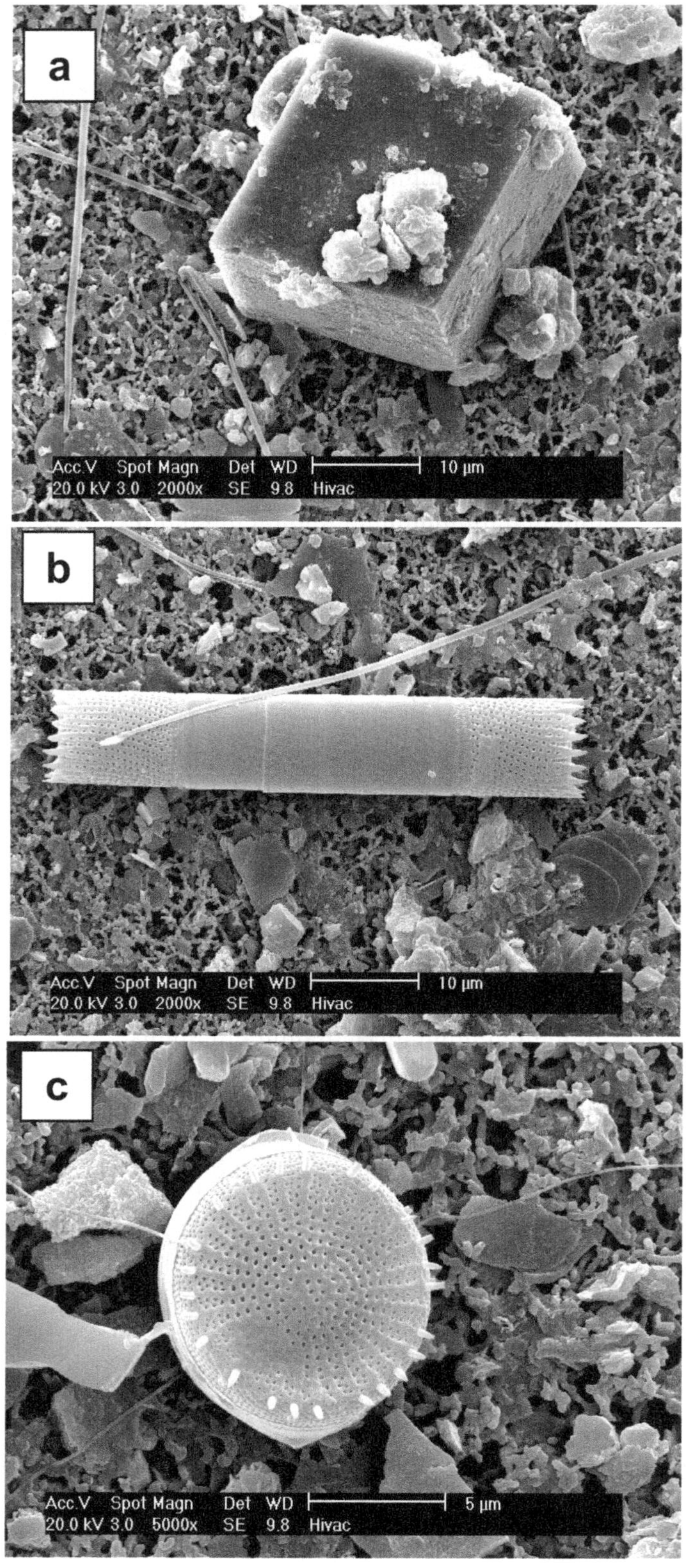

Figure 4.48 SEM image of an (a) iron crystal, (b) *Cyclotella* (d = 10 μm), and (c) *Melosira* (d = 3 -12 μm) on a 0.45 μm filter from the filter influent at the Salem, Oregon water treatment facility.

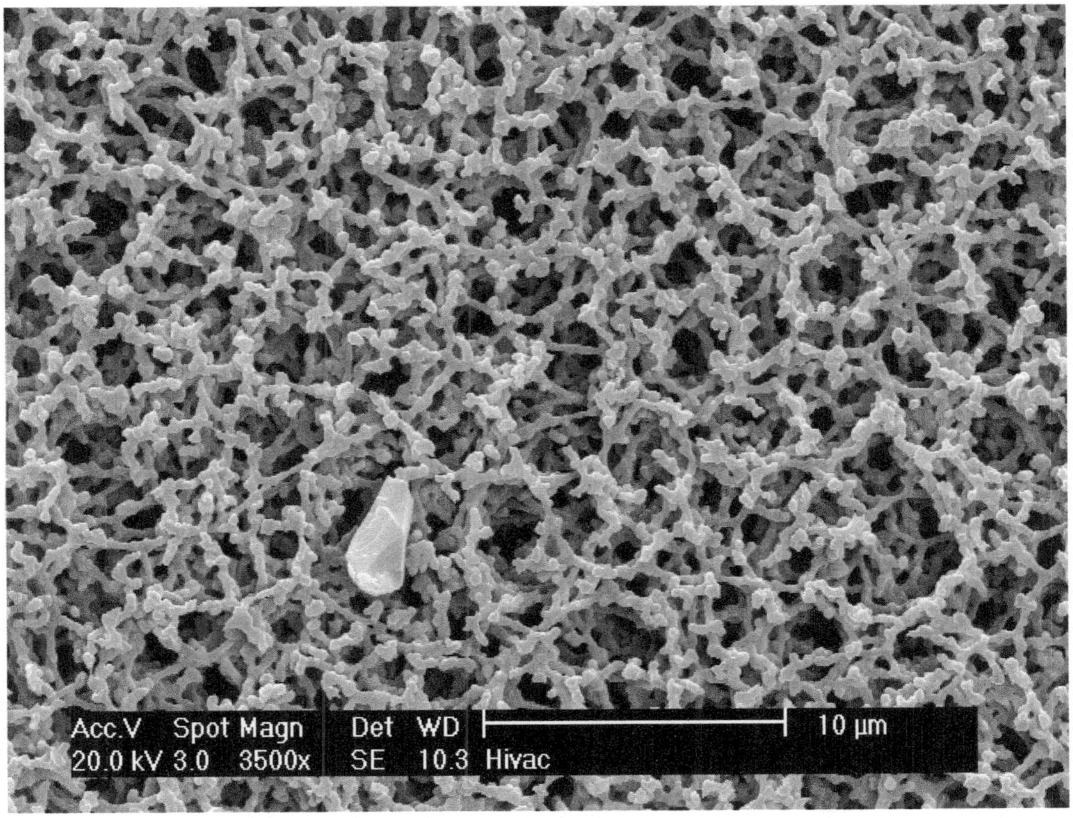

Figure 4.49. SEM image of a 0.45 µm filter-pad deposit from the effluent from Filter #4 at the Salem, Oregon water treatment facility.

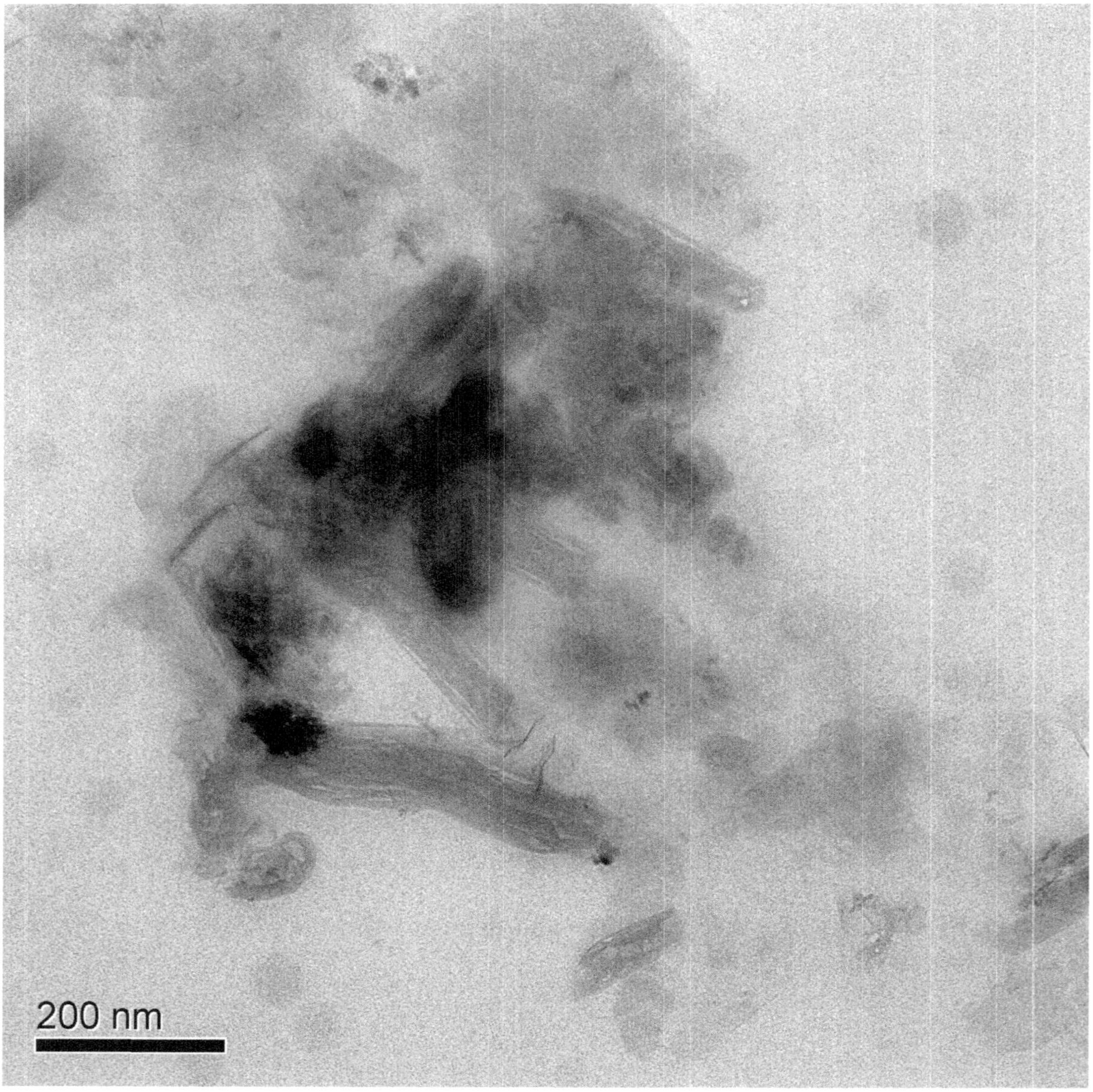

Figure 4.50 TEM image of ultrafine particles deposited on a copper mesh microgrid from the influent to Filter #4 at the Salem water treatment facility

CHAPTER 5
COMPREHENSIVE ANALYSIS

In this chapter a comprehensive discussion of the results from all of the participating plants is presented. The results for each of the treatment facilities reported in previous chapter are compared with respect to process characteristics and water quality. The effect of various parameters such as filter influent water quality, pretreatment processes employed, and filter operation and design on the characteristics of materials entering and exiting the filters are evaluated.

COMPOSITION

Overall, the inorganic and organic constituents in the filter influent of various utilities were predominantly present in the ultrafine and dissolved + size fractions, respectively (Figure 5.1a and b). This was observed during both stressed and unstressed operating conditions. For each size fraction the standard deviation was significant (coefficient of variation (CV) > 0.4) indicating the high variability of water sources and pretreatment among the different facilities.

The relative size distributions of inorganic and organic materials in the filter effluent during stressed and unstressed conditions are reported in Figure 5.2. It should be noted that the mass fraction reported on the y-axis is plotted on a log scale to better illustrate small percentages of material present in the settleable and filtrable size classes. Both the inorganic and organic materials were predominantly present in the ultrafine and dissolved + size fractions during both stressed and unstressed conditions. This is expected since the filter influent also had a similar distribution of particle sizes. The total inorganic mass remained relatively unchanged from the filter influent to the effluent at approximately 46 ppm during the stressed conditions and 43 ppm during the unstressed conditions. Similarly, the mean effluent organic concentrations remained relatively unchanged at 2.9 and 2.4 ppm during the stressed and unstressed periods, respectively.

The filtrable fraction of inorganic material increased with filter run time (i.e., from start to end) and was coupled with an associated decrease in the amount of materials in the settleable fraction. A similar trend was observed for both stressed and unstressed conditions. This indicates that the larger materials are being removed by the filter as filtration progressed and ripening occurred.

The bulk of organic material was largely present in the dissolved + size fraction, followed by ultrafine, settleable, and filtrable in decreasing order of organic material presence. It should be noted that the percentage of organic material present in the settleable size range is significantly higher than the inorganic material present in this size range. Then again the percentage of inorganic material present in the dissolved + and ultrafine size ranges is higher than the organic material present in these size ranges.

A TEM image of some ultrafine particles is shown in Figure 5.3. The particles shown here are approximately 50 nm in diameter indicating that they are between the ultrafine and dissolved + size fractions.

For all of the utilities the inorganic material in the influent and effluent was primarily composed of calcium, magnesium, and silica, which accounted for 98% of the total concentration of approximately 50 ppm (Table 5.1). Iron and aluminum were the most prevalent elements in the remaining fraction for both filter influent and effluent. Iron and aluminum in the influent may have originated from a variety of sources including that found naturally in the raw water, though the bulk of it likely originated from the addition of the aluminum and iron coagulants in the coagulation and flocculation processes. The concentrations of both aluminum and iron varied significantly and can be attributed to a variety of factors including differences in coagulant dosage and type used, coagulation conditions such as pH, temperature, and the use of coagulant aids (e.g., polyaluminum chloride (PACl)), in addition to differences in the raw water variability at various plants. Of all the utilities studied only Northern Kentucky and Southern Nevada used an iron-based coagulant (i.e., ferric chloride) while the others used aluminum sulfate (alum). Northern Kentucky however, used PACl as a coagulant aid while Southern Nevada did not. Consequently, Northern Kentucky had an influent aluminum concentration to the filters of approximately 0.2 ppm while the concentration in the filter influent for Nevada was negligible.

The mean concentration of aluminum in the filter influent was approximately 0.4 ppm during stressed and unstressed conditions. The effluent concentration of aluminum was approximately 0.1 ppm during the stressed operating condition and 0.4 ppm during the unstressed condition. The higher concentration of aluminum in the effluent during the unstressed condition could be due to the relatively higher concentration observed at the Milwaukee Water Works facility, compared to the other facilities. This could be attributed to the specific operating conditions and raw water quality during the sampling event that might have led to slight deviation from optimal treatment conditions. For instance, at the Milwaukee Water Works facility, the aluminum concentration in the filter influent during the unstressed sampling period was 1.7 ppm, while during the stressed period, the concentration was 0.7 ppm. Exclusion of Milwaukee's data resulted in similar mean effluent concentrations of aluminum during both stressed and unstressed conditions.

The concentration of iron in the filter influent during stressed and unstressed conditions was similar at approximately 0.18 and 0.16 ppm, respectively. The effluent concentrations during these two operating conditions were lower. The effluent concentrations during stressed and unstressed conditions were 0.04 and 0.11 ppm, respectively.

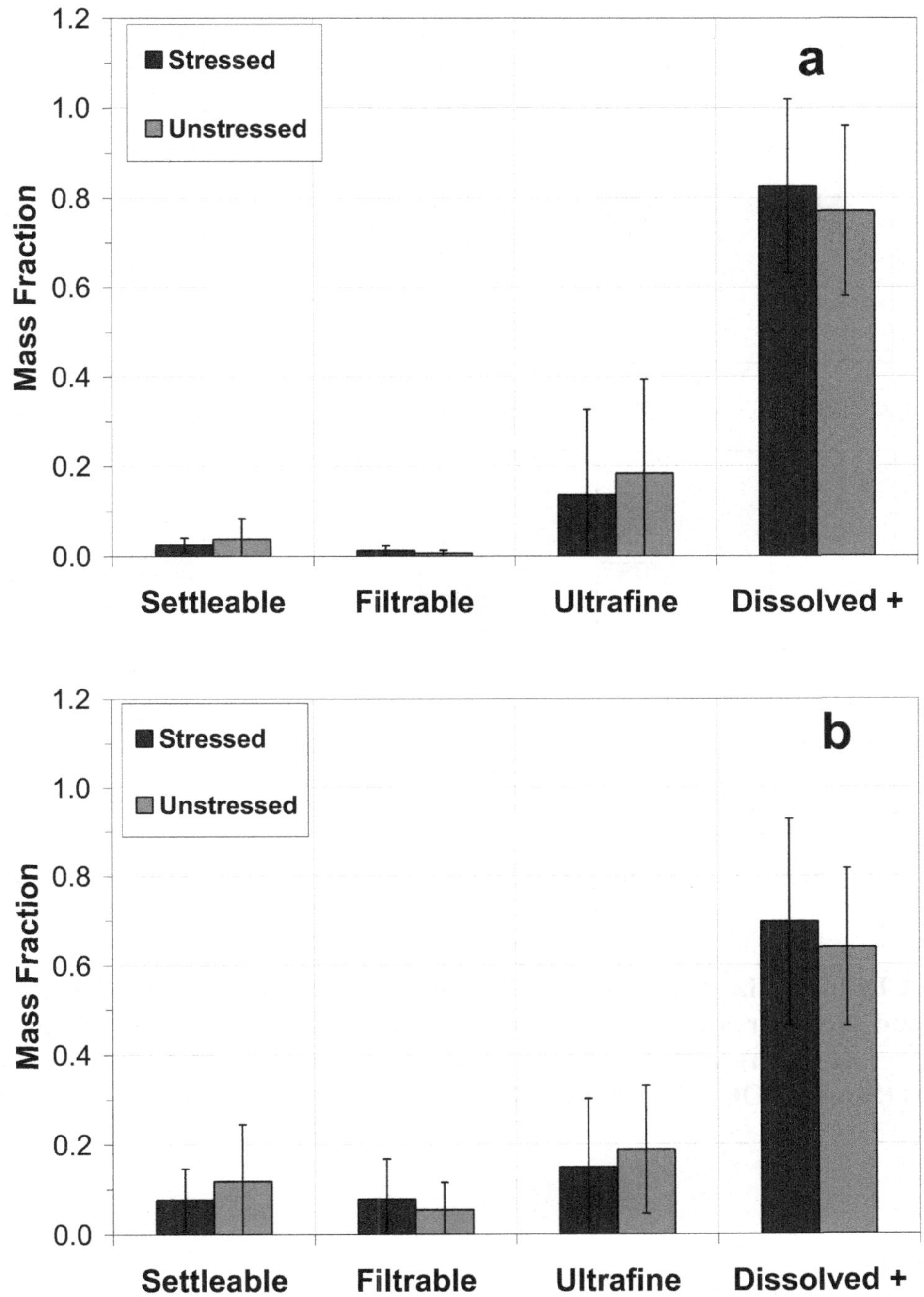

Figure 5.1 Mean relative size distribution of (a) inorganic and (b) organic materials in the filter influent. Error bars represent the standard deviation of the mean value (Stressed $[\text{Metals}]_T$ = 55 ppm; Unstressed $[\text{Metals}]_T$ = 44 ppm; Stressed $[\text{TOC}]_T$ = 2.98 ppm; Unstressed $[\text{TOC}]_T$ = 2.81 ppm)

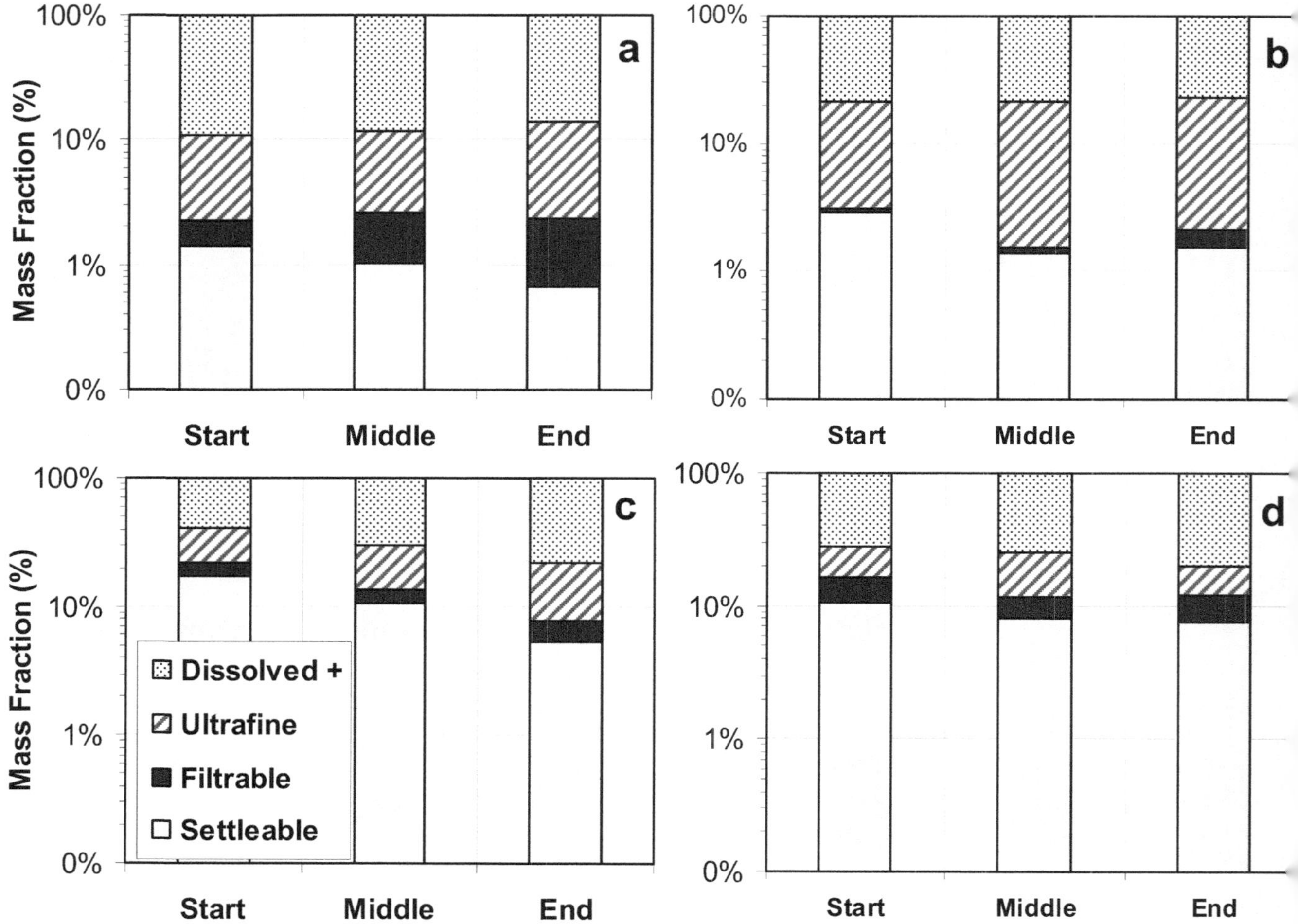

Figure 5.2 Effluent size fractions as a function of filter run time for inorganics under (a) stressed and (b) unstressed operating conditions (Stressed $[Metals]_T$ = 53 ppm; Unstressed $[Metals]_T$ = 42 ppm) and for TOC under (c) stressed and (d) unstressed operating conditions (Stress $[TOC]_T$ = 2.98 ppm; Unstressed $[TOC]_T$ = 2.81 ppm)

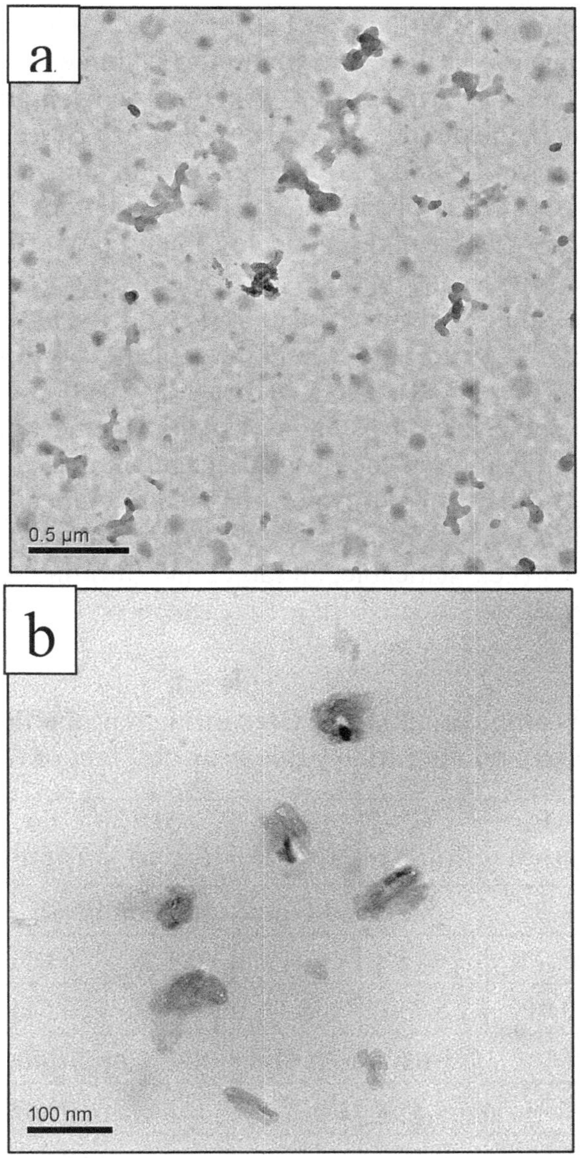

Figure 5.3 TEM picture of effluent ultrafine particles deposited by ultracentrifugation from (a) filter 10 and (b) 16 at the Milwaukee Water Works treatment plant

The distribution of coagulant residuals in the influent and effluent across the four size fractions (i.e., settleable, filtrable, ultrafine, and dissolved +) was determined and is reported in Figure 5.4. The size distributions of aluminum residual in the influent and effluent are shown in Figures 5.4a 5.4b, while that of iron are shown in Figures 5.4c and 5.4d. Aluminum in the influent was nearly equally distributed across the filtrable (26%), settleable (35%), and dissolved + (35%) size fractions, with a smaller amount present in the ultrafine fraction (6%) (Figure 5.4a). A substantial fraction of aluminum in the filter effluent is present in the dissolved + size class (67%), followed by settleable (16%), ultrafine (11%) and filtrable (6%). The distribution of aluminum into various size classes can be attributed to the presence of aluminum as aluminum precipitates, and dissolved aluminum that might be bound to organic ligands and inorganic colloids in various size classes. With respect to the concentration of the aluminum in various size classes, significant reduction in the concentration of aluminum in the settleable (80% removal) and filtrable (91% removal) size classes was observed. Approximately 20% removal was observed for aluminum in the dissolved + and ultrafine size classes.

Results for iron were similar to aluminum with iron largely present in the settleable (18%), filtrable (58%), and dissolved + (23%) size fractions in the filter influent for those utilities using iron based coagulants (Figure 5.4c). In the filter effluent approximately 95% of the influent iron in the settleable and filtrable size fractions was removed. However, iron that was present in the ultrafine and dissolved + fractions was relatively unaffected by filtration and was only reduced by about 2%. As a result, iron in the effluent was primarily present in the dissolved + (86%) size fraction (Figure 5.4d). The actual size distribution for both the aluminum and iron species in the influent and effluent was highly variable amongst the different utilities (CV = 100 to 200%). Thus, the distribution of the residual coagulant materials appears to be dependent on the unique conditions at the different water treatment facilities (e.g., pH, temperature, organic levels, alkalinity, and treatment processes used). The presence of residual coagulant species in the larger size fractions (i.e., settleable, filtrable, and ultrafine) indicates that they have either become bound to other, larger materials or are present as precipitates.

Table 5.1
Mean elemental composition of the filter influent and effluent during stressed and unstressed operating periods at the start of the filter run

	Ca (ppm)	Mg (ppm)	Si (ppm)	Fe (ppm)	Al (ppm)	Pb (ppm)	Cu (ppm)	As (ppm)	Mn (ppm)
Stressed Operating Condition									
Influent	30.26	10.11	6.15	0.179	0.425	0.015	0.001	0.0032	0.012
Effluent	29.83	9.91	5.75	0.038	0.099	0.014	0.001	0.00044	0.005
Unstressed Operating Condition									
Influent	30.80	7.90	4.83	0.159	0.375	0.007	0	0.00033	0.005
Effluent	29.84	6.69	4.66	0.111	0.388	0.011	0	0.00017	0.0004

Concentrations of lead, copper, arsenic, and manganese in the filter influent and effluent were significantly lower than those of the other elements that were analyzed (Table 5.1). The mean concentration of lead in the filter effluent during the stressed period for all of the utilities was equal to or less than the action level (0.015 ppm) set by the USEPA. Copper was present at a mean concentration of approximately 1 part per billion ppb in the influent and effluent during stressed conditions, while its concentration was negligible during the unstressed period. The mean concentration for arsenic in the filter influent was approximately 3 and 0.3 ppb during the stressed and unstressed operating periods, respectively. These concentrations were reduced in the effluent streams to approximately 0.4 ppb and 0.17 ppb during the same respective periods. However, there was a significant amount of variability (CV ~ 100%) in the arsenic concentrations for both the influent and effluent streams (Table 5.1). For example, the highest arsenic concentration was measured at the Salem utility where the influent concentration was 24 ppb. However, the effluent concentration for all utilities was 3 ppb or less, well below the proposed 10 ppb standard for arsenic in drinking water.

The types of inorganic materials retained within the filter were further determined through an analysis of the filter backwash following each filter run. In general, the hardness causing cations, calcium and magnesium, made up the bulk mass in the backwash, accounting on average, for 60% of the total mass of inorganics that were analyzed. Aluminum however, was also a large constituent (~20 ppm or 20%) in the average filter backwash water. Recalling that each utility, except Southern Nevada, utilizes alum as their primary coagulant, the aluminum in the effluent is due to the residual coagulant from the sedimentation basins. However, in contrast to the results for aluminum, lower concentration of iron ([Fe] = 0.03 ppm) was present in the backwash water for either of the utilities studied here. This is surprising given that the mean influent iron concentration at this facility was higher than that measured for aluminum ([Fe] ~ 0.5 ppm, [Al] ~ 0.2 ppm). The lack of iron in the backwash for Northern Kentucky may be attributed to the fact that the backwash water must first be filtered through a 0.45 μm filter prior to ICP analysis. Thus, a large fraction of iron and aluminum species may be removed from the backwash water prior to ICP analysis. Nevertheless, aluminum accounts for a significant portion of the inorganic material that remains in the filter. Furthermore, based on the analysis of the influent, effluent, and backwash waters, the granular media filters are effective in removing residual coagulant species. However, they do not appear to significantly reduce the concentration of these materials that are in the dissolved + size fraction, which is where the bulk of the residual coagulant resides (Figure 5.4).

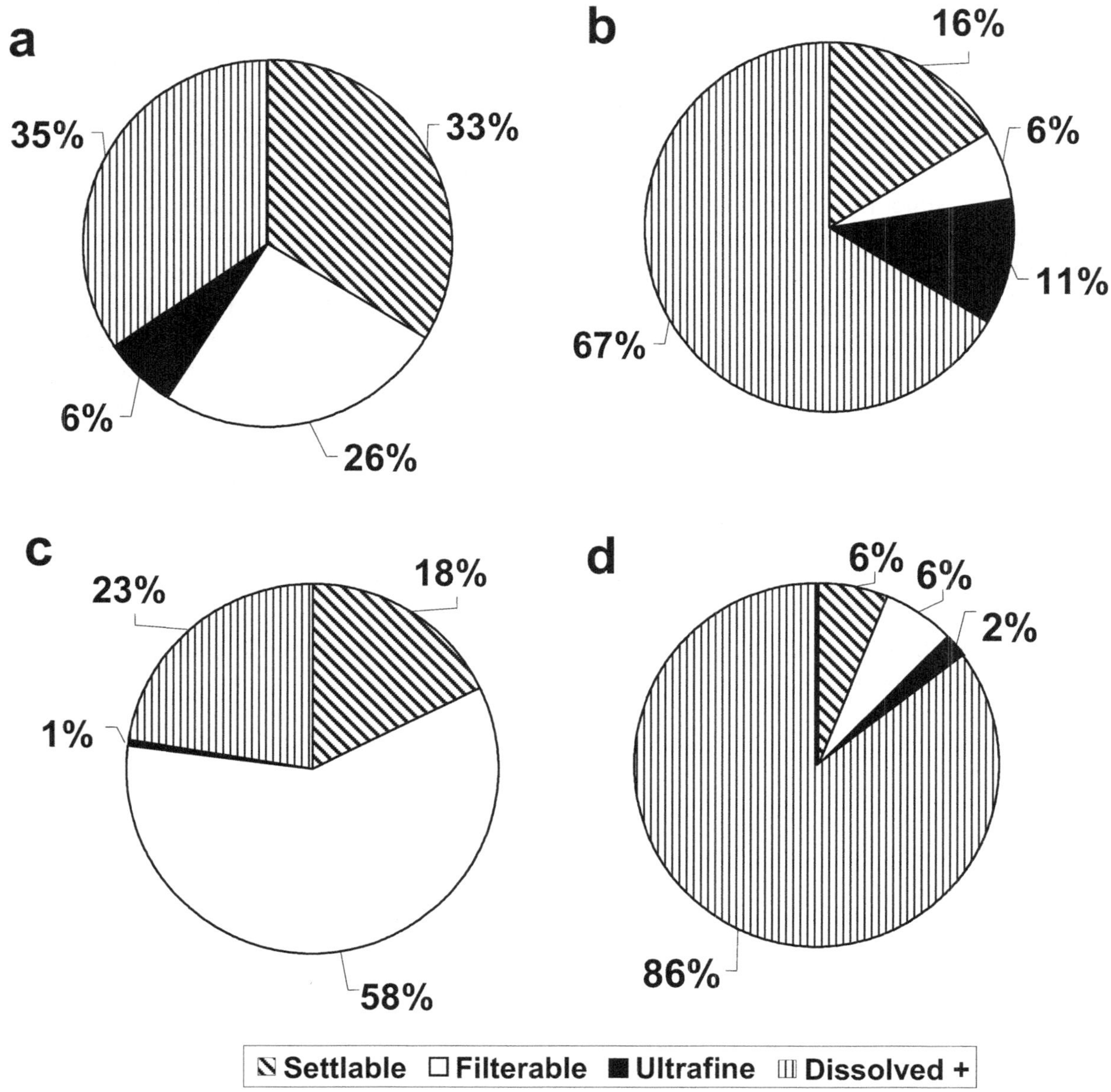

Figure 5.4 Average size distribution of aluminum in the filter (a) influent and (b) effluent and for iron in the filter (c) influent and (d) effluent for the water treatment facilities using aluminum and iron based coagulants (mean influent $[Al]_T = 0.437$ ppm; mean effluent $[Al]_T = 0.183$ ppm; mean influent $[Fe]_T = 0.396$ ppm; mean influent $[Fe]_T = 0.109$ ppm). Results were calculated considering both stressed and unstressed conditions together

Figure 5.5 shows an SEM image of a filter cake retained on a 0.45 μm filter pad following filtration of the filter influent and effluent from the Milwaukee Water Works plant. From Figure 5.5a it is evident that the influent had a considerable amount of material that was retained on the 0.45 μm filter resulting in the formation of a cake layer. EDAX analysis of this cake revealed that is was mostly composed of aluminum (94%), calcium (4%), and some iron (2%). In comparison little filtrable material was detected in most cases for the filter effluent (Figure 5.5b). This type of cake in the influent and the absence thereof in the effluent was seen for the majority of the treatment plants participating in this study. Exceptions to this trend were observed for the filter effluent at the Northern Kentucky (Figure 5.6) and Buffalo Pound utilities. At the Northern Kentucky plant the effluent cake layer was largely composed of silica (50%), with smaller amounts of aluminum (10%) and iron (10%). Similarly at the Buffalo Pound plant the effluent cake was composed mostly of silica and aluminum in addition of other organic constituents. The spherical objects in the effluent cake varied in size and were made up of aluminum (~25%), iron (~10%), and silica (~55%) along with some sodium, calcium, and manganese. The amount of filtrable material in the effluent for Northern Kentucky varied with the duration of filter run and which filter was sampled. For instance, for two of the three filters that were sampled, the amount of filtrable material in the effluent was highest at the start and middle of the filter run while for the third filter it was nonexistent at the start and middle but present at the end of the filter run. This suggests that for the first two filters removal increased as the filter ripened while for the third filter some breakthrough may be occurring at the end of the filter run. Nevertheless, the presence of these materials in the filter effluent suggests that its operation could be further optimized. It is possible that some degree of complexation occurred between the residual coagulants and the silica to produce the filtrable material in the effluent in these cases. However, given that other utilities had higher effluent concentrations of aluminum and comparable amounts of organic and inorganic materials (e.g., Milwaukee) the formation of the filtrable materials is due to unique conditions at the respective utilities.

In addition to the inorganic materials that were retained on the 0.45 μm filter, various forms of biota (Figure 5.7) were also detected. Such constituents in the filtrable fraction were diverse. At the Salem Water Treatment Plant a significant number of large filter-clogging diatoms (up to 100 μm in size) were identified under the SEM, as *Fragilaria*, *Cyclotella*, and *Melosira* (APHA, 1994 #1044). It is likely that most of these diatoms originated from sediment accumulated in the raw water intake pipe. The diatoms may serve as transport vehicles for harmful microorganisms that would be shielded during the downstream disinfection processes and may signify less than ideal filter performance. It is possible that the presence of these diatoms in the effluent would not be detected by conventional turbidity measurements due to the transparent nature of the cell walls of these microorganisms. Therefore, additional detection tools may be required for those facilities in which these diatoms are present.

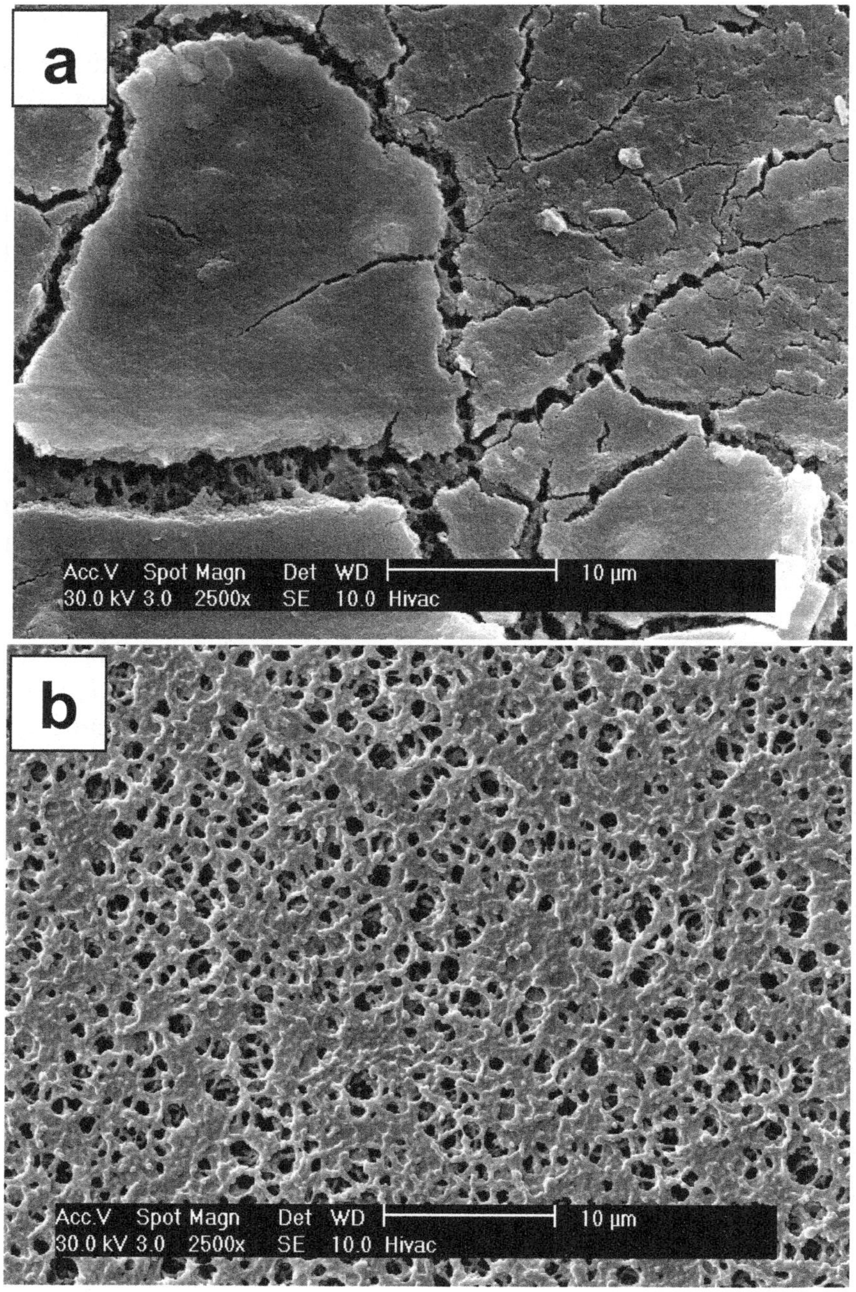

Figure 5.5 SEM image of filter cake after filtration of the Milwaukee Water Works (a) influent and (b) effluent during the unstressed operating period. The filter has a pore size of 0.45 μm and is a cellulose nitrate material

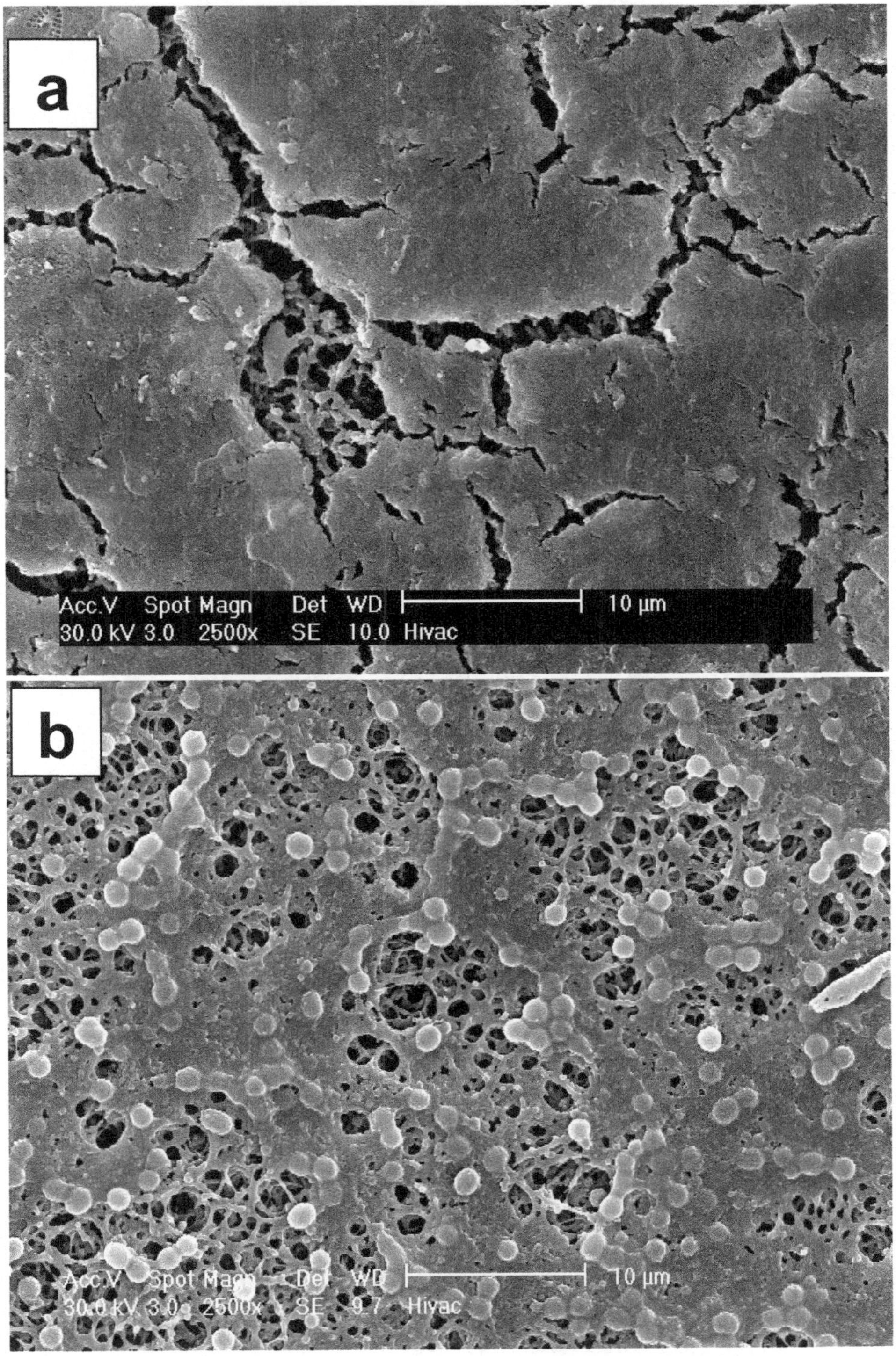

Figure 5.6 SEM image of the filter cake after filtration of the Northern Kentucky (a) influent and (b) effluent during the unstressed operating period. The filter has a pore size of 0.45 μm and is a cellulose nitrate material

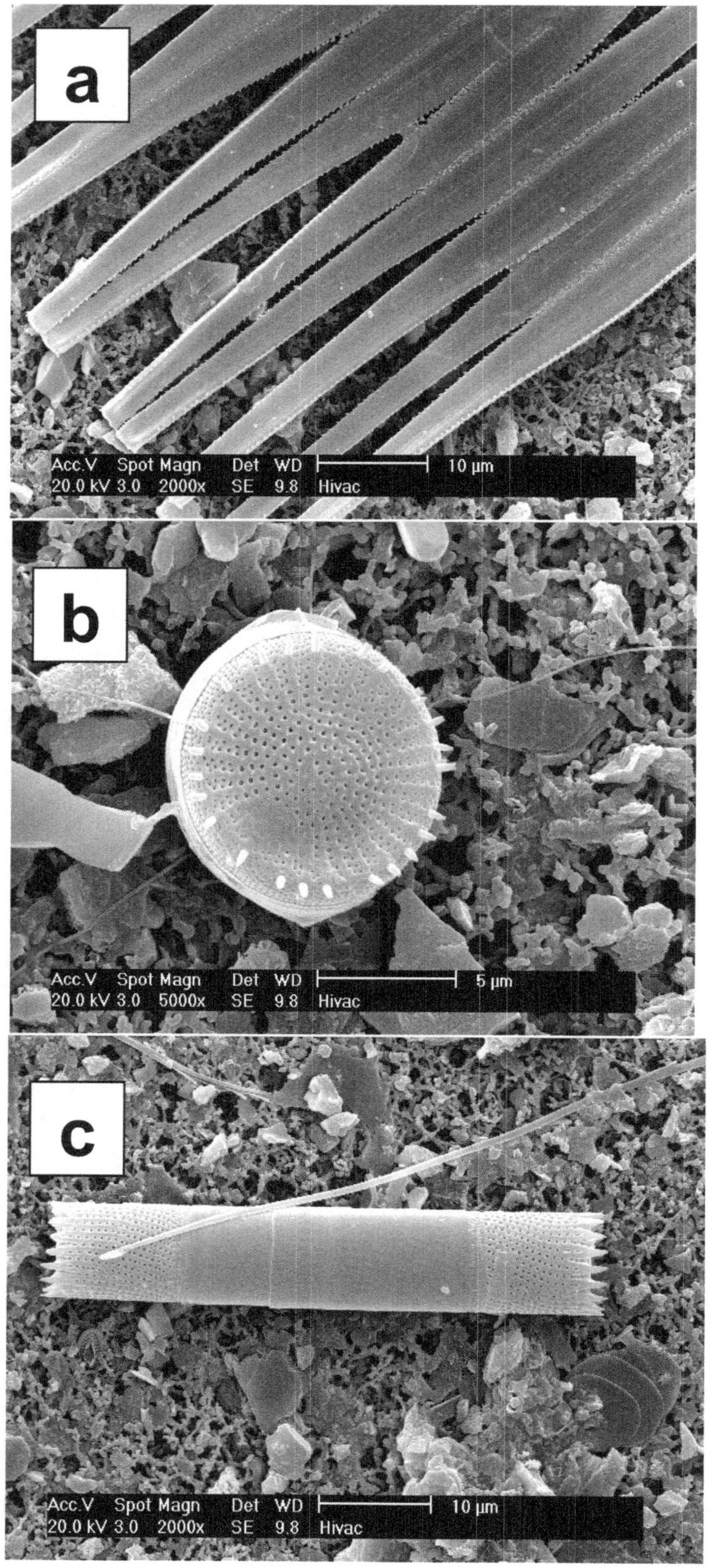

Figure 5.7 SEM images of (a) *Fragilaria* (100 µm), (b) *Cyclotella* (10 µm), and (c) *Melosira* (3 × 12 µm) found in the filter effluent of several water treatment plants

OPERATIONAL FACTORS AND FILTER PERFORMANCE

In this section, a comparative evaluation of the results from all plants is performed. The results are evaluated based on differences in pretreatment methods employed, pre-oxidants, the type(s) of coagulants and coagulant aids used during pretreatment, source water chemistry and the filter operating parameters such as filtration rate and filter media type (Chapter 3).

Effect of Pretreatment

Pre-oxidation

The purposes of pre-oxidation are oxidation of organic compounds into harmless forms, taste and odor control, iron and manganese oxidation into insoluble forms and color control. Generally, common oxidants used in water treatment are chlorine, ozone, chlorine dioxide, potassium permanganate and hydrogen peroxide. However, if the organic matter content of the raw water is high, then harmful disinfection by-products may form, particularly when chlorine is used.

All of the participating utilities use some type of pre-oxidant; therefore it was difficult to evaluate the differences between the filter influent characteristics with and without pre-oxidation. Results were therefore evaluated based on the effect of pre-oxidant type on filter performance and particle characteristics. The relative size distribution of particles in the filter influent for all utilities during stressed and unstressed conditions are shown in Figure 5.8 and 5.9, respectively. In most cases, the dissolved + size fraction was the predominant size fraction in both the stressed and unstressed periods. However, at a number of facilities more mass was present in the larger size fractions during the unstressed period compared to that in the stressed period.

For Modesto and Milwaukee, which use pre-ozonation, greater than 50% of the total organic mass was in the larger size fractions (i.e., settleable, filtrable or ultrafine) during the unstressed period (Figure 5.9). However, under stressed conditions the bulk of the TOC mass was in the dissolved + size fraction (Figure 5.8). Similarly, inorganics in the filter influent at Modesto and Milwaukee were present mostly in the larger size fractions during unstressed conditions. The biggest difference between the two plants was the relative size of the ultrafine fraction, which remained small at Milwaukee plant during the unstressed period. As noted earlier, particles may aggregate in the filter influent through complexation reactions between calcium and organic materials and/or through increased association of aluminum with ozonated organic matter. Pre-ozonation may increase the degree of complexation of the particles. The effect of pre-ozonation may be influenced by the influent hardness to TOC ratio. Chang and Singer (1988) reported that water with low hardness to TOC ratios were less susceptible to the effects of ozone than waters with hardness to TOC ratios of 20 or higher. This ratio was about 6-7 for Modesto and 38-48 for Milwaukee during the stressed and unstressed conditions, respectively. Thus, the impact of pre-ozonation on the particle size distribution may be less for Modesto due to the low hardness to TOC ratio for the raw water at this facility. Another plant, which uses pre-ozonation, was the Southern Nevada plant. However, at this facility the trend was not similar to that of Modesto and Milwaukee, with almost all the mass present in the dissolved + size range. This may be because of the different coagulant, ferric chloride, used at this facility, while alum is used at the other two facilities. The majority of metal species was dissolved while ultrafine TOC increased at Southern Nevada. As a result, effect of pre-ozonation on the larger size particle fractions was not observed at the Nevada plant.

Pre-chlorination was employed at the Buffalo Pound utility. Relatively large organic and inorganic particle size fractions were also observed at the Buffalo Pound utility compared to other utilities. Approximately 55% and 40% of the organic and inorganic particles, respectively, were in the large sized particle size fractions (settleable and ultrafine) during the stressed period. This may be due to the effect of pre-chlorination that occurred prior to filtration. For instance, Huck et al. (2002) observed that chlorine is the most efficient pre-oxidant for destabilizing particles to induce aggregation. The filter run time was relatively short at the Buffalo Pound plant, which may be explained by the high ultrafine fraction in the filter influent.

Chlorine dioxide and potassium permanganate were used as pre-oxidants at Johnson County. The filter influent at Johnson County was observed to have the most substantial fraction of inorganic and organic materials in the larger size fractions (Figure 5.8), mostly in the ultrafine size class during stressed period. This may be due to the impact of chlorine dioxide and/or potassium permanganate on the relative particle size distribution.

Potassium permanganate was the only pre-oxidant used at the Cincinnati, Northern Kentucky and Houston facilities. The amount of particles in the larger size fractions is higher at the Cincinnati and Houston plants during unstressed periods. While the observed results may be attributed to the use of potassium permanganate at these two facilities, the absence of materials in the larger size fractions at the Northern Kentucky facility during both stressed and unstressed conditions indicates that the potassium permanganate is not inducing particle aggregation in the filter influent in this case. These results suggest that the effect of permanganate on mass size distribution should be considered along with other factors such as water quality, coagulant used and operating conditions.

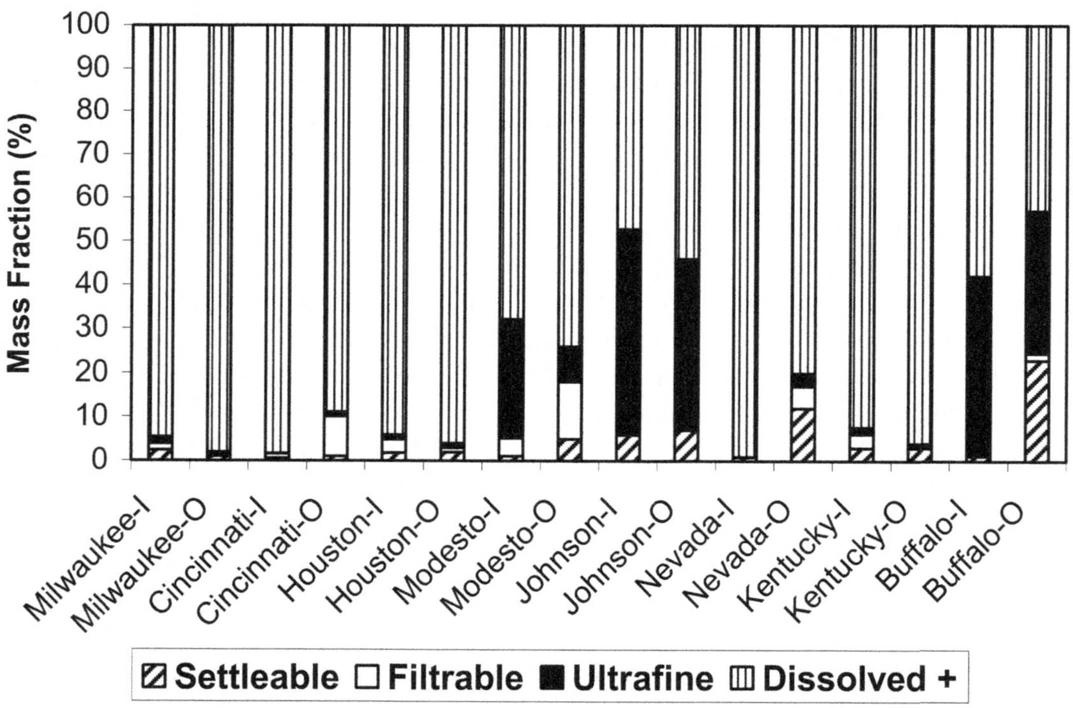

Figure 5.8 Organic and inorganic particle size distribution of all filter influents at the investigated utilities during the stressed period (I = inorganic and O = organic)

Oxygen, permanganate, chlorine dioxide, ozone, hydrogen peroxide and free chlorine are all capable of oxidizing iron and manganese. The choice of the oxidant for iron and manganese oxidation depends on several parameters such as efficiency of oxidation, required dosage and presence of other constituents that exert demand on oxidants (e.g., sulfide, nitride, ammonia, dissolved organic carbon) (Langlais, Reckhow, and Brnk 1991). Due to the extremely fast kinetics of iron oxidation, uncomplexed Fe(II) can easily out compete Mn(II) for the oxidant. Therefore, little manganese oxidation occurs at insufficient doses and at relatively high concentrations of iron.

Iron and manganese concentrations in the filter influent and effluent during the middle of filter runs are given in Tables 5.2 and 5.3, respectively for selected participating facilities. Generally, the iron was present in the larger size fractions in the filter influent, and was subsequently removed during filtration. However, iron was also present in the ultrafine and dissolved + fractions in the Milwaukee, Modesto and Kentucky filter influent water and these utilities also had higher total iron concentrations in the filter influent. Except Milwaukee, the iron in the influent in the larger size classes is almost completely removed. The removal of iron in the dissolved + size class ranged from 2 to 25%. For Milwaukee, most of the iron present in the influent was in the ultrafine and dissolved + size classes. However, in the filter effluent, iron was present primarily in the filtrable and dissolved + size classes. The data indicates that the iron removed from the dissolved + and ultrafine size classes aggregated into larger sized particles that contributed to the increased percentage of mass in the filtrable size range.

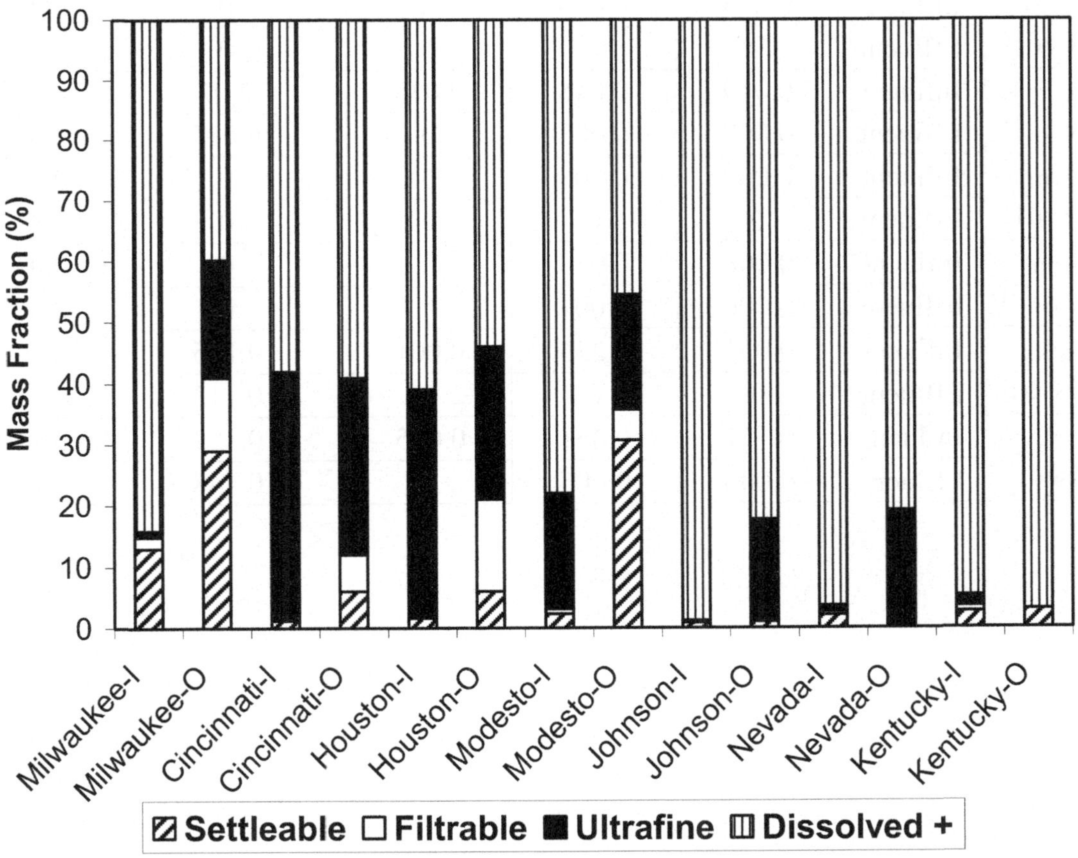

Figure 5.9 Organic and inorganic particle size distribution of all filter influents at the investigated utilities during the unstressed period (I = inorganic and O = organic)

Manganese concentrations were low in all the utilities. It was either present in the dissolved + size range or in filtrable and settleable size ranges. Manganese present in the filtrable and settleable size classes in the influent was typically removed completely. Manganese, when present in high concentrations (for example 0.045 mg/L in the influent of Cincinnati and Buffalo Pound), it was completely removed. When present in lower concentrations, removal ranged from none to 80%.

Table 5.2
Concentration of iron in the different size classes in the filter influent and effluent for selected water treatment utilities. The reported total concentration is the sum of the mean concentrations reported for each size class. The results are reported for the middle of the filter run.

		Concentration (ppm)				
		Settleable	**Filtrable**	**Ultrafine**	**Dissolved +**	**Total**
Milwaukee (Unstressed)	Influent	0.005	0.038	0.119	0.355	0.517
	Effluent	0.008	0.288	0	0.167	0.463
Cincinnati (Stressed)	Influent	0.086	0.193	0	0	0.279
	Effluent	0	0	0	0	0
Houston (Stressed)	Influent	0.014	0.056	0	0	0.07
	Effluent	0	0	0	0	0
Modesto (Stressed)	Influent	0.019	0.046	0.033	0.02	0.118
	Effluent	0.007	0.007	0	0.015	0.029
Johnson (Stressed)	Influent	0.067	0.008	0	0	0.075
	Effluent	0	0	0	0	0
Nevada (Unstressed)	Influent	0.008	0.033	0	0	0.041
	Effluent	0.006	0.001	0	0	0.007
Kentucky (Unstressed)	Influent	0.051	0.228	0.001	0.115	0.395
	Effluent	0	0	0	0.113	0.113
Kentucky (Stressed)	Influent	0.151	0.441	0.005	0.152	0.749
	Effluent	0.006	0.005	0	0.148	0.159

Table 5.3
Concentration of manganese in the different size classifications in the filter influent and effluent for selected water treatment utilities. The reported total concentration is the sum of the mean concentrations reported for each size class. The results are reported for the middle of the filter run.

		Concentration (ppm)				
		Settleable	Filtrable	Ultrafine	Dissolved +	Total
Milwaukee (Stressed)	Influent	0	0	0	0.001	0.001
	Effluent	0	0	0	0.001	0.001
Cincinnati (Unstressed)	Influent	0.04	0.005	0	0	0.045
	Effluent	0	0	0	0	0
Houston (Stressed)	Influent	0.002	0.001	0	0.009	0.012
	Effluent	0.001	0	0	0.009	0.01
Modesto (Stressed)	Influent	0	0.003	0	0.002	0.005
	Effluent	0	0	0	0.001	0.001
Buffalo (Stressed)	Influent	0.04	0.005	0	0	0.045
	Effluent	0	0	0	0	0
Kentucky (Unstressed)	Influent	0	0	0	0.003	0.003
	Effluent	0	0	0	0.002	0.002

Coagulant Residuals and Filter Performance

The overriding objective of this investigation was to comprehensively characterize the material that enters and exits water treatment filters. To this end, residual coagulant species have been identified as a potentially significant source of material entering water treatment filters, which may impair filtered water quality and filter performance. The coagulant residuals themselves may clog the filter media (for example, precipitates of aluminum and iron salts not removed in the clarification process), or may lead to increased concentrations of soluble metals. For instance, elevated residual coagulant concentrations may result in discoloration and sediment deposition in the distribution system. In most cases aluminum or iron salts are used as a primary coagulant because neither presents any substantial health effects under normal circumstances. The target concentrations for aluminum and iron in the finished water are less than 0.2 and 0.3 ppm (CFR 2002), respectively. The coagulation process should be optimized to minimize residual concentrations of the coagulants used.

The participating utilities in this study predominantly use aluminum sulfate (alum) as their primary coagulant (Table 3.1). The exceptions to this are the Northern Kentucky which uses a blend of alum and polymer as primary coagulant and ferric sulfate as coagulant aid and Southern Nevada which uses ferric chloride as as the primary coagulant. Johnson County, and Northern Kentucky use polyaluminum chloride (PACl) as a coagulant aid. Cincinnati, Modesto, and Northern Kentucky also use cationic polymers as flocculant aids.

The concentration of residual coagulant, either aluminum or iron, in the filter influent and effluent was determined and the results are separately reported in Figures 5.10a and b, respectively. The reported values represent the mean concentration for the respective metals from all of those utilities using the given metal salt as their primary coagulant. Additionally, the results are differentiated according to operating condition (i.e., stressed and unstressed). For those plants using alum as their primary coagulant the average filter influent aluminum concentrations were 0.4 and 0.15 ppm during the stressed and unstressed periods, respectively (Figure 5.10a). Effluent aluminum concentrations were close to 0.05 ppm, and were slightly higher during the stressed period compared to the unstressed one. The concentrations of aluminum in the effluent did not change considerably with filter run time although a gradual decrease was observed as the filter run progressed. On average, about 74% of the total residual aluminum concentration in the filter influent was removed. Removal of the aluminum by the filter is demonstrated by an analysis of the filter backwash water during both the stressed and unstressed periods where the mean aluminum concentrations were 18 and 10 ppm, respectively. Notably, the higher backwash concentration of aluminum during the stressed period corresponded to higher influent concentrations of aluminum during this period (Figure 5.10a). The aluminum concentration in the backwash water varied significantly among the utilities. For example, the highest concentration was measured at Buffalo Pound ([Al] = 66 ppm) and the lowest at Johnson County ([Al] = 0.04 ppm). This variability may be attributed to a myriad of factors including coagulant dosing, filter type, and the concentration of organics in the filter influent. Results for the influent and effluent aluminum concentrations at the Milwaukee facility during the unstressed period were omitted from Figure 5.10a as they were considerably higher than all other values and would have skewed the reported results, as discussed earlier. At this facility the influent aluminum concentration was 1.8 ppm with effluent concentrations at the start, middle, and end of the filter run of 1.7, 2.3, and 3.0 ppm, respectively.

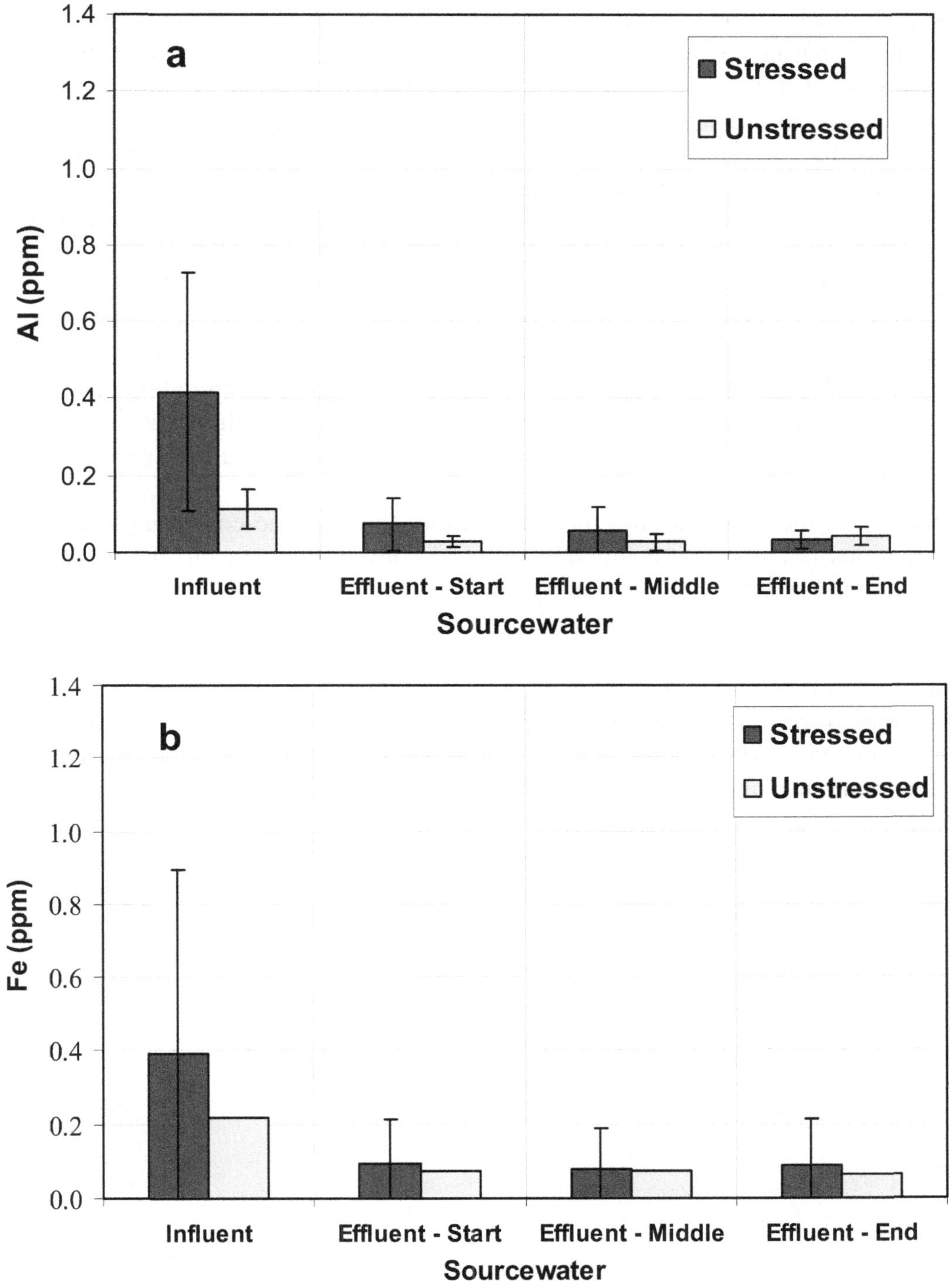

Figure 5.10 Bulk concentrations of (a) aluminum and (b) iron in the filter influent and effluent for those water treatment facilities using alum and iron based coagulants, respectively. Data is reported for both the stressed and unstressed operating conditions. Error bars represent the standard deviation of the mean reported values

In contrast to the other utilities previously discussed, both the Northern Kentucky and Southern Nevada utilities use an iron based salt (ferric chloride) as their primary coagulant. The mean concentration of iron in the filter influent was approximately 0.4 and 0.2 ppm for the stressed and unstressed periods, respectively (Figure 5.10b). Effluent concentrations of iron were considerably lower, at approximately 0.1 ppm, resulting in an average removal of around 74%. As was observed for the aluminum, the effluent iron concentration did not change considerably with filter run time. However, in contrast to the results for aluminum, low concentrations of iron ([Fe] = 0.03 ppm) were present in the backwash water for both utilities. This is surprising given that the mean influent iron concentrations were higher than that measured for aluminum (Figure 5.10). At Northern Kentucky, which uses ferric sulfate as a coagulant aid with a blend of alum and polymer as the primary coagulant, the concentration of iron was low in the backwash ([Fe] = 0.04 ppm) compared to that of aluminum ([Al] = 17 ppm). This would suggest that iron might remain adsorbed to the filter media more strongly than does the aluminum and is not as effectively removed during backwashing. However, given the fact that the backwash water had to be filtered prior to analysis, removing some of the materials in the sample makes this point speculative.

In Figure 5.11, the concentration of aluminum and iron in the combined settleable and filtrable size fractions is plotted as a function of TOC concentration in the filter influent and effluent. Similarly, in Figure 5.12, the concentration of aluminum and iron in the dissolved + size fraction is plotted as a function of dissolved organic carbon concentration in both the filter influent and effluent. In both cases, no correlation between TOC and coagulant residual concentrations in various size fractions was observed.

The presence of residual coagulant in the filter influent may contribute to increased filter clogging and associated headloss. Figure 5.13 shows filter run time as a function of the filter loading rate for aluminum and iron for the different utilities. No strong correlation for either coagulant could be found to suggest that lower concentrations of residual coagulant species in the filter influent results in longer filter run times.

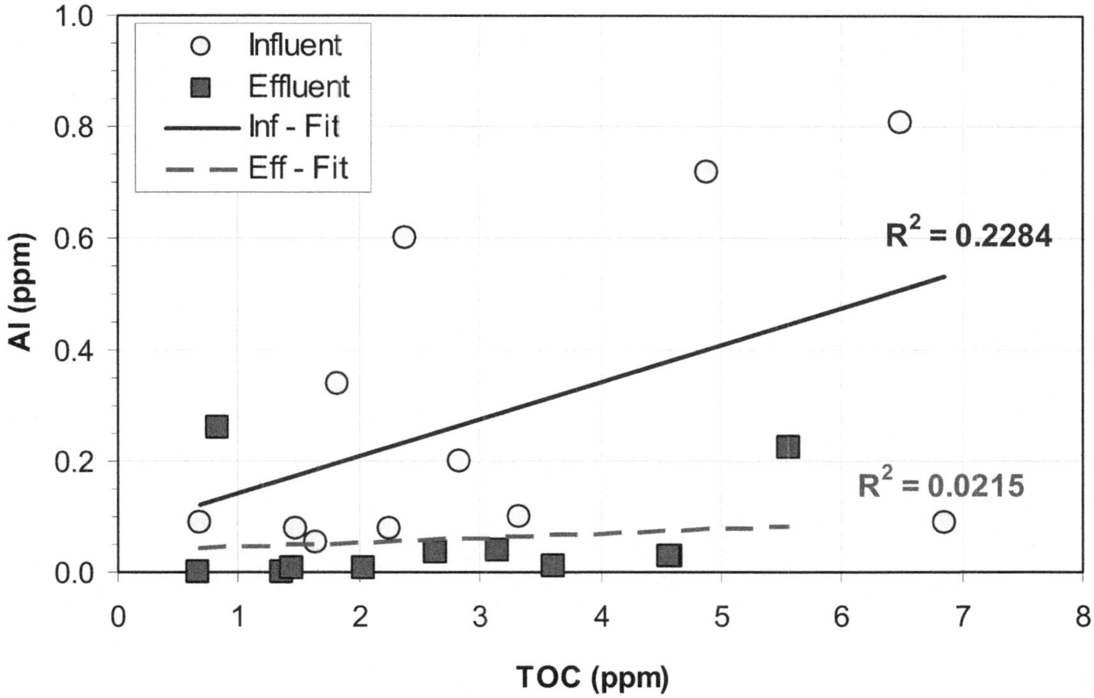

Figure 5.11 Concentration of (a) aluminum and (b) iron in the combined settleable and filtrable size fractions in the filter influent and effluent as a function of TOC concentration

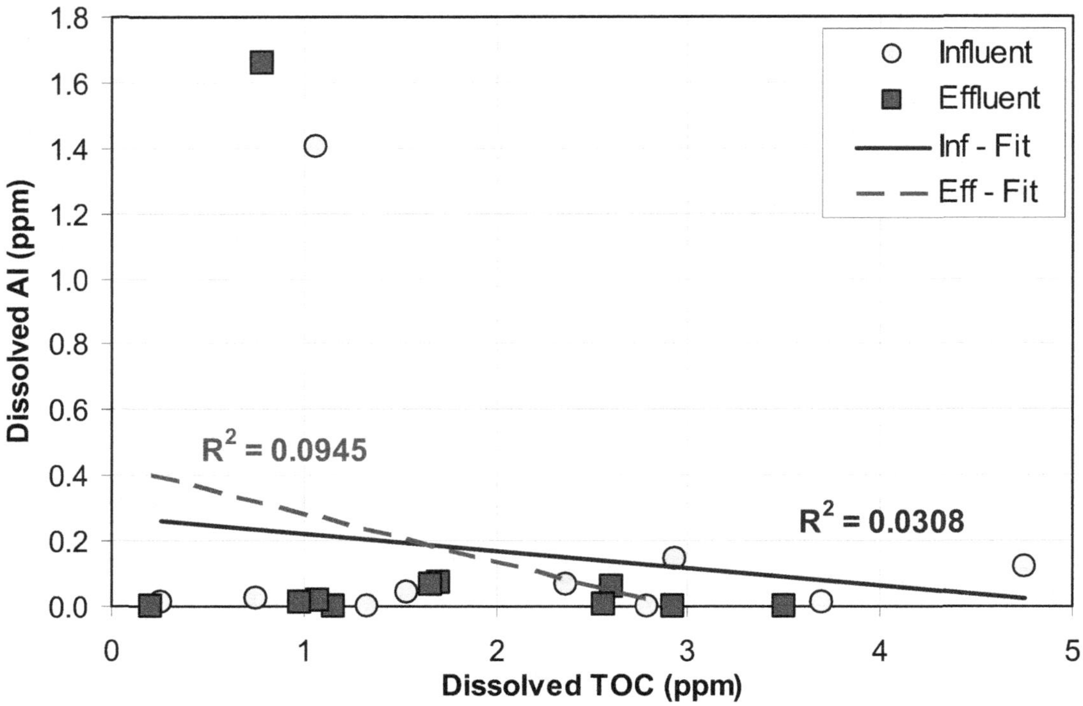

Figure 5.12 Concentration of (a) aluminum and (b) iron in the dissolved + size fraction in the filter influent and effluent as a function of dissolved TOC concentration

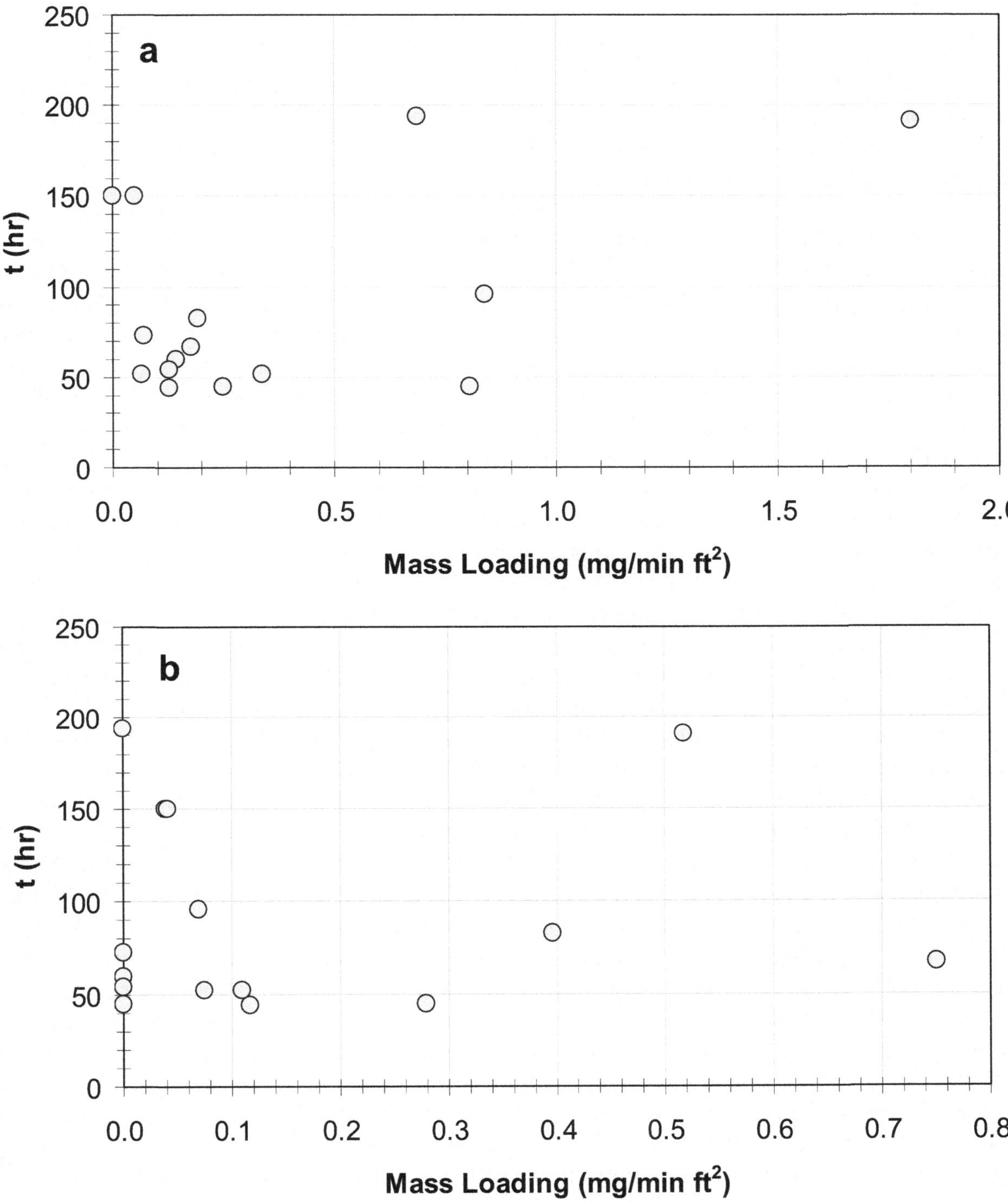

Figure 5.13 Filter run-time for each of the participating utilities as a function of the loading rate for (a) aluminum and (b) iron to the filter beds

Filter Operational Factors

Filtration Rate

Filtration rate directly impacts the loading rate of various materials on to the filter media and also influences the magnitude of shear forces acting on retained particles. Therefore, filtration rate can have significant effects on the attachment to and detachment of particles from the filter media. For example, as filtration rate increases previously retained materials may be sheared off and exit the media through the filter effluent.

The filtration rates used at the participating utilities ranged from approximately 2 to 6 gpm/ft^2. The Buffalo Pound, Milwaukee, Northern Kentucky and Cincinnati utilities operate at filtration rates of 2.55, 2.75, 2.85 and 3 gpm/ft^2, respectively. The Southern Nevada, Houston and Modesto utilities operate at a substantially higher filtration rate of 6 gpm/ft^2, while Johnson County uses a filtration rate of 5 gpm/ft^2. In this section, the effect of low (e.g., Milwaukee, Cincinnati, Buffalo Pound, and Northern Kentucky) and high filtration rates (e.g., Modesto, Johnson County, Southern Nevada and Houston) are compared to the size distribution of both organic and inorganic materials. Influent and effluent total concentrations and associated mass fractions for the different groups are given in Tables 5.4 and 5.5, respectively. Additionally, the total mass fraction of larger size particle fractions (settleable, filtrable, and ultrafine) for inorganic material in the filter influent and effluent are given in Figures 5.14, 5.15, 5.16 and 5.17 during the unstressed and stressed periods. It was expected that the removal of organic material in the larger size classes (settleable, filtrable and ultrafine) in the effluent would be higher at lower filtration rates (~ 2.55 - 3 gpm/ft^2) compared to that at the higher filtration rates (~ 5 - 6 gpm/ft^2). Such a trend was not observed consistently. For instance, the organic mass in the larger size classes in the filter effluent at Milwaukee was lower compared to the influent during unstressed periods. However, during stressed conditions, the concentration in the effluent was higher than the influent, suggesting creation of organic particles in the larger size class within the filter. Similarly for Northern Kentucky, it appears that larger sized organic particles were created within the filter, during both stressed and unstressed periods. Hence, the interactions within the filter media appear to have a more significant effect than the filter loading rate on the concentration of larger organic particles in the effluent. On the other hand, the effluent concentrations of inorganic particles in the larger size classes were either lower or unchanged compared to those in the influent for facilities operating at lower filtrate rates. At higher filtration rates, some facilities exhibited good removal of these larger inorganic particles, while others did not. In addition, the stage of filter run affected the concentration of larger particles. For instance, the effluent concentration of larger inorganic particles in the effluent of Johnson County's filter was higher than that of the influent, suggesting potential breakthrough of these particles. The increase in concentration of inorganic material in the larger size fractions correlated with an increase in calcium and magnesium concentrations suggesting that these particles could be precipitates of calcium and magnesium. These data suggest that the concentration of larger inorganic particles in the effluent may be affected to a greater degree than that of organic particles. Particularly, the potential for generation of organic particles in the larger size classes within the filter media may mask any effects of filtration rate on the removal of the particles in these size classes.

Filter Media Type

Filter media type determines the surface area that is available for particle deposition and retention (Crittenden, et al. 2005). Furthermore, the hydrodynamic characteristics (i.e., porosity and flow channel characteristics) of the filter are also dependent upon the type of media used. Selection of filter media involves a trade-off between headloss (larger media minimizes headloss) and filtration efficiency (smaller media captures particles better) (Crittenden, et al. 2005).

There are two types of filter media used in the treatment facilities studied here: fine sand and anthracite coal. Notably the Modesto and Cincinnati water treatment plants use only anthracite coal and sand, respectively, as a filter media while the other utilities have dual media filters composed of sand and anthracite coal (Table 3.1). The diameter of the anthracite particles is larger than that for sand, resulting in comparatively larger pores within the media. For this reason, the removal efficiency for the dual media filters is expected to be high, especially for larger particles. The removal percentage of larger organic particles by the monomedia filters at Cincinnati is observed to be lower than that of other filters operating at similar loading rates. However, the removal of larger inorganic particles was higher at Cincinnati than other facilities with similar loading rates. While the reason for this is not clear, it may be attributable to a higher shearing rate that would occur in the small pores of the sand layer in multimedia filters. It appears that inorganic deposits may be more susceptible to shearing than organic deposits. The removal rate observed in Modesto of the larger organic and inorganic particles did not show any such clear trend compared to other multimedia filters that also operated at similar loading rates. It appears that at high loading rates, media alone may not affect filter performance significantly.

Table 5.4
Total concentration of organic and inorganic materials in the settleable, filtrable, and ultrafine size fractions at low and high filtration rates (I = inorganic; O = organic)

Plants	Filtration Rate (gpm/ft^2)	Influent		Start		Middle		End	
		O (ppm)	I (ppm)	O (ppm)	I (ppm)	O (ppm)	I (ppm)	O (ppm)	I (ppm)
Unstressed									
Milwaukee	2.75	0.76	9.58	0.03	1.86	0.05	1.03	0.03	3.83
Kentucky	2.85	0.05	2.16	0.57	0.01	0.54	0.35	0.60	0.04
Cincinnati	3	0.92	32.44	0.88	33.95	1.18	27.95	1.33	32.10
Stressed									
Buffalo	2.55	3.70	32.98	2.65	33.13	2.65	26.96	2.98	31.47
Milwaukee	2.75	0.02	2.75	1.48	1.28	1.54	1.01	1.46	0.93
Kentucky	2.85	0.07	4.02	0.54	2.54	0.45	2.08	0.49	3.76
Cincinnati	3	0.36	0.91	0.82	1.96	0.17	0.73	0.33	0.18
Unstressed									
Johnson	5	0.39	0.51	0.00	0.06	0.00	0.14	0.00	3.15
Houston	6	3.16	25.4	1.10	23.24	1.83	20.43	1.83	19.5
Modesto	6	0.90	2.63	0.46	1.50	0.45	1.46	0.41	1.51
Nevada	6	0.59	3.06	1.22	3.15	0.59	3.64	0.41	3.2
Stressed									
Johnson	5	1.30	27.82	0.95	31.14	0.86	32.60	0.94	35.15
Houston	6	0.15	3.41	1.97	5.60	0.94	5.97	0.38	3.59
Modesto	6	0.38	3.98	0.27	1.73	0.29	1.62	0.30	1.32
Nevada	6	0.63	1.14	0.21	2.47	0.00	2.80	0.00	2.17

Table 5.5

Mass fraction of organic and inorganic materials in the combined settleable, filtrable, and ultrafine size groups at low and high filtration rates (I = inorganic; O = organic)

Plants	Filtration Rate (gpm/ft²)	Influent		Start		Middle		End	
		O (%)	I (%)	O (%)	I (%)	O (%)	I (%)	O (%)	I (%)
Unstressed									
Milwaukee	2.75	42	17.3	4	3.6	6	2.1	4	7.3
Kentucky	2.85	4	5.3	34	0	33	0.9	35.4	0.1
Cincinnati	3	41	42	43	43.1	47.6	35.6	52.5	41.3
Stressed									
Buffalo	2.55	57	42.1	47.4	43.1	46.4	35.6	52.5	41.3
Milwaukee	2.75	2	5.4	47	2.7	48	2.1	45	1.8
Kentucky	2.85	4	7.7	33	5.1	27	4.1	29	7
Cincinnati	3	11.1	1.7	29	3.7	8	1.4	14	0.3
Unstressed									
Johnson	5	17.7	1.2	0.0	0.1	0.0	0.3	0.0	7.1
Houston	6	46	39.1	24	38.9	38	34.3	38	32.3
Modesto	6	54.5	22	32	13.9	30	13.5	28.8	13.9
Nevada	6	19.2	3.6	33	2.7	17.2	3.1	12	2.7
Stressed									
Johnson	5	46	52.8	36	59.5	32	51.9	37	64.9
Houston	6	4	5.9	43.4	10.2	26.8	10.8	13	6.9
Modesto	6	26	32.1	20.1	14.5	21	13.5	22.2	11.1
Nevada	6	20	1	7	2	0	2	0	2

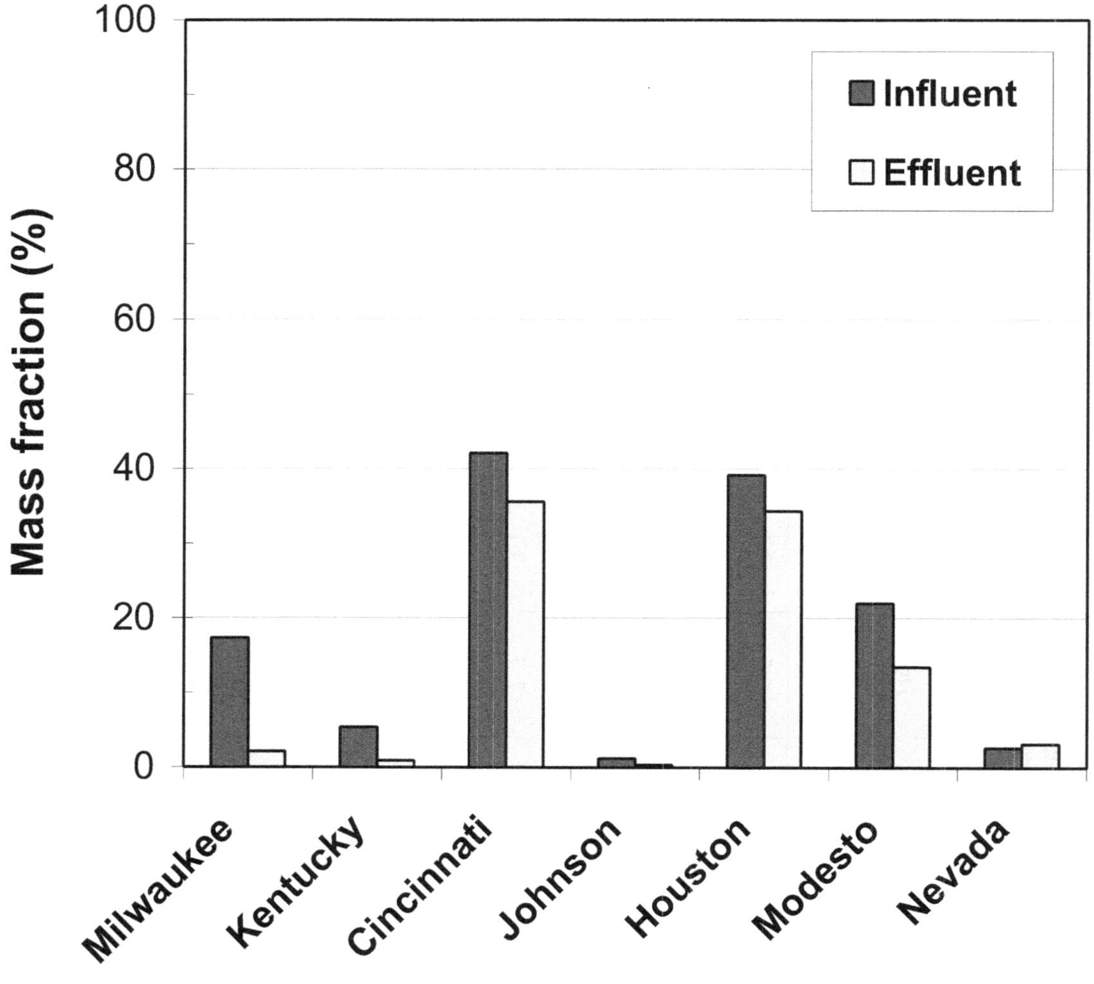

Figure 5.14 Mass fraction of inorganic particles in the larger size classes (settleable, filtrable, and ultrafine) during the unstressed period (Filtration rates for Milwaukee, Northern Kentucky and Cincinnati are 2.75, 2.85 and 3 gpm/ft^2 and filtration rates for Johnson, Houston, Modesto and Southern Nevada are 5, 6, 6 and 6 gpm/ft^2, respectively)

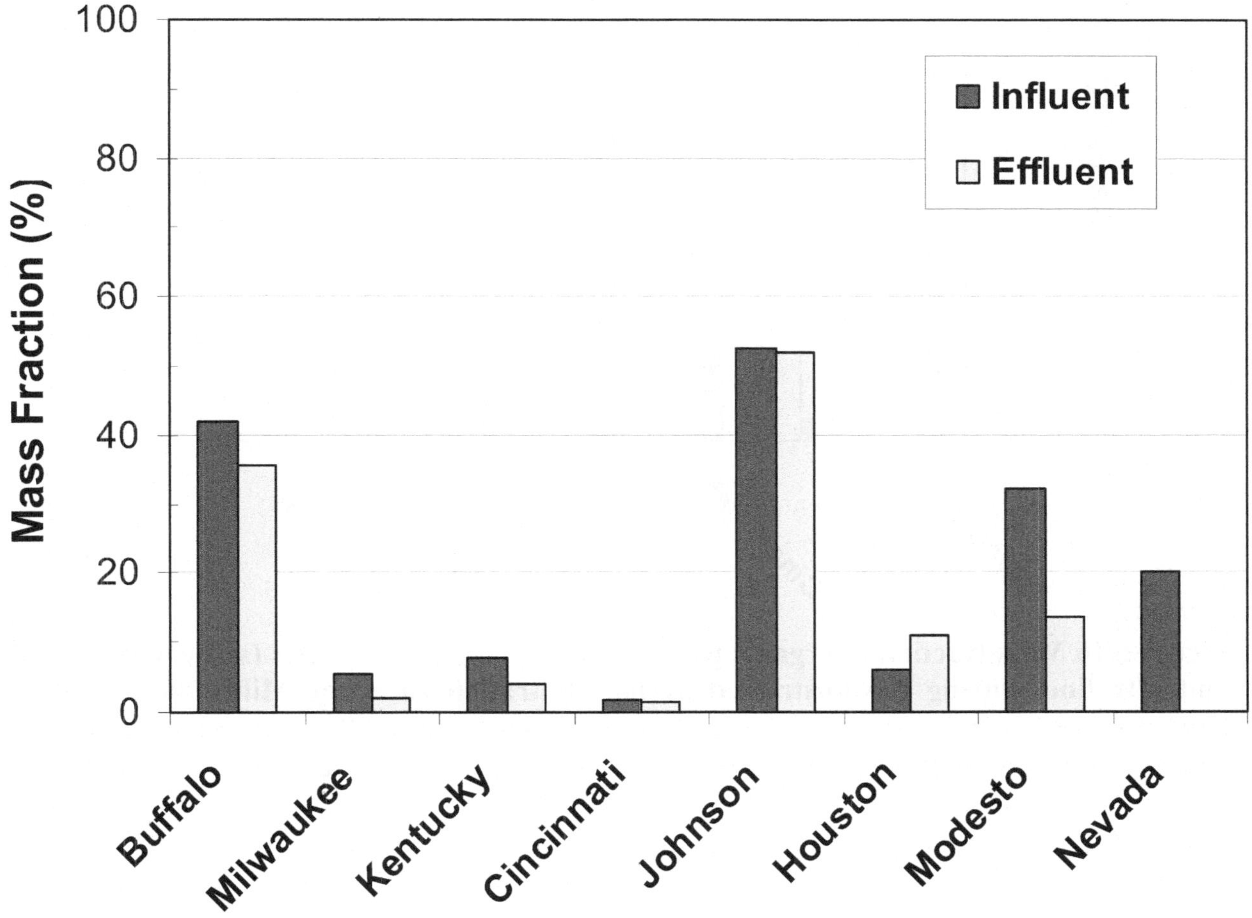

Figure 5.15 Mass fraction of inorganic particles in the larger size classes (settleable, filtrable, and ultrafine) during the stressed period (Filtration rates for Buffalo Pound, Milwaukee, Northern Kentucky and Cincinnati are 2.55, 2.75, 2.85 and 3 gpm/ft^2 and filtration rates for Johnson County, Houston, Modesto and Southern Nevada are 5, 6, 6 and 6 gpm/ft^2, respectively)

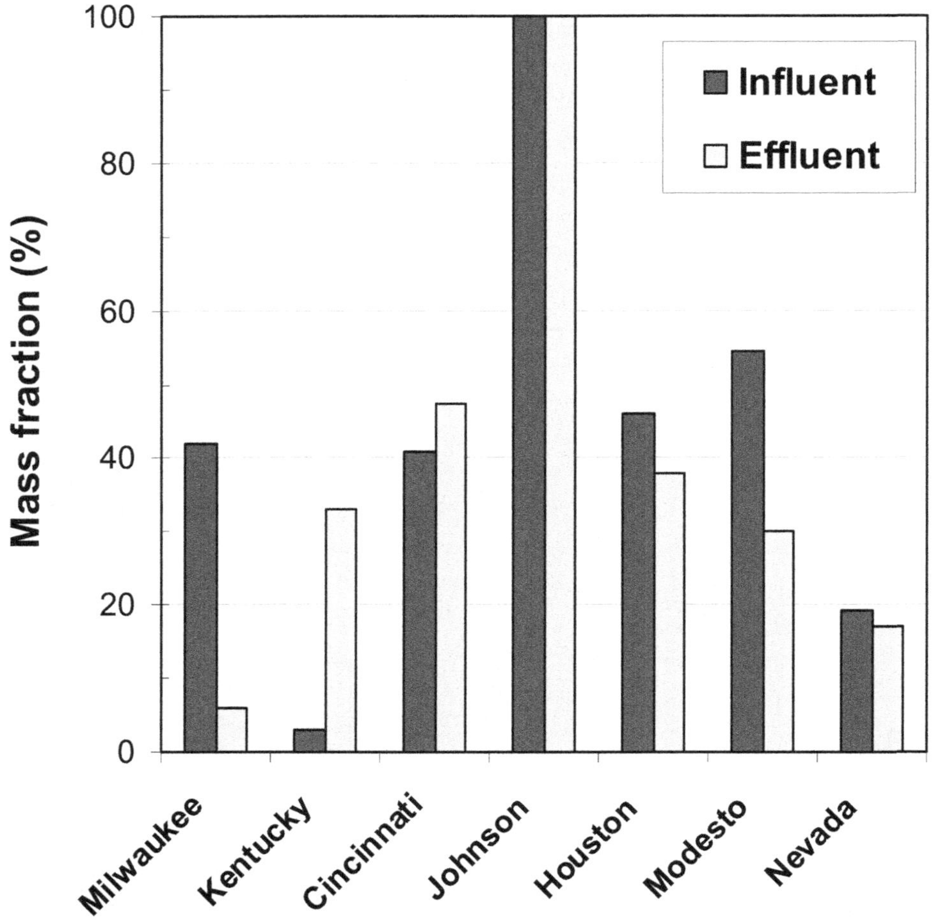

Figure 5.16 Mass fraction of organic particles in the larger size classes (settleable, filtrable, and ultrafine) during the unstressed period (Filtration rates for Milwaukee, Northern Kentucky and Cincinnati are 2.75, 2.85 and 3 gpm/ft^2 and filtration rates for Johnson County, Houston, Modesto and Southern Nevada are 5, 6, 6 and 6 gpm/ft^2, respectively)

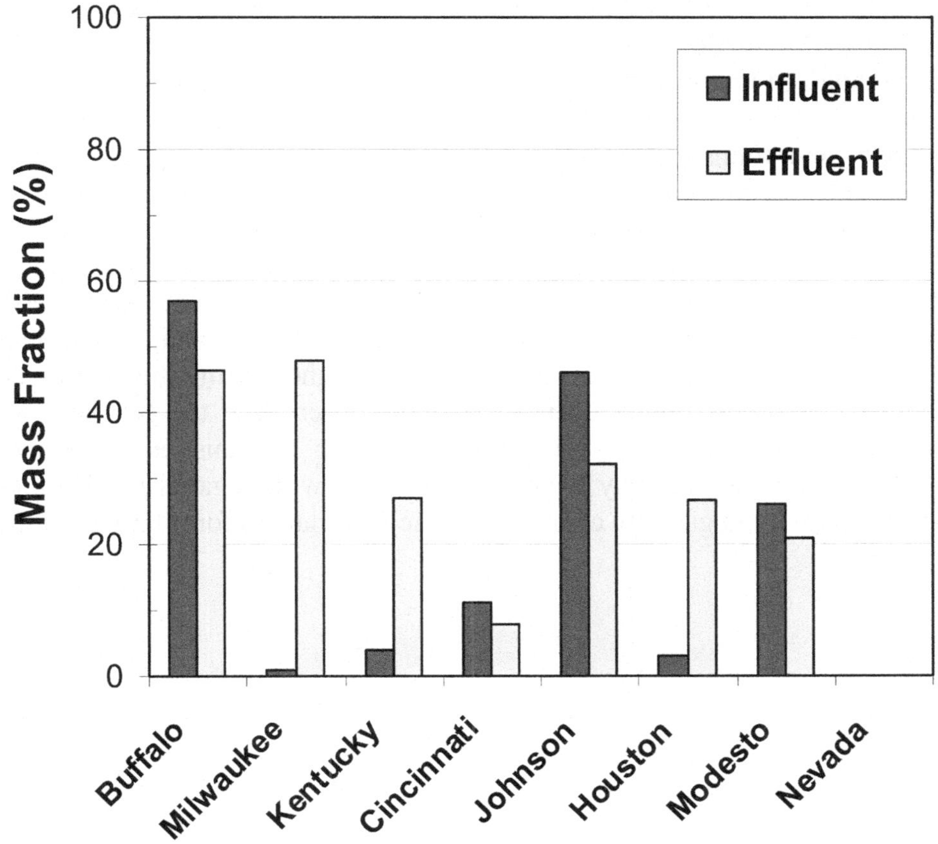

Figure 5.17 Mass fraction of organic particles in the larger size classes (settleable, filtrable, and ultrafine) during the stressed operating period. The filtration rates for Buffalo Pound, Milwaukee, Northern Kentucky and Cincinnati are 2.55, 2.75, 2.85 and 3 gpm/ft^2, respectively. The filtration rates for Johnson County, Houston, Modesto and Southern Nevada are 5, 6, 6 and 6 gpm/ft^2, respectively

Stressed and Unstressed Periods

As noted in Chapter 3, filter influent and effluent at several facilities were evaluated during stressed or unstressed periods that were determined based on historical data of water quality and filter performance. In the following section we compare the quality, as determined by the concentrations of organic carbon and selected inorganic species in the water, of the filter influent during these two conditions. The selected water quality results for the filter influent during the stressed and unstressed conditions at the participating utilities are given in Table 5.6.

Generally the concentrations of organic material in the filter influent were higher during the stressed periods compared to that measured during the unstressed periods. In two cases, Houston and Modesto, the organic carbon concentrations were higher in the influent during the unstressed period relative to the stressed period. For Northern Kentucky, however, the organic carbon concentration did not change between the stressed and unstressed periods. The elevated organic carbon concentrations were not however, associated with increased residual coagulant (i.e., Fe and Al) concentrations. In fact, higher residual coagulant concentrations were measured during periods of lower organic carbon concentrations in several cases (e.g., Houston, Milwaukee, and Cincinnati). Otherwise the residual coagulant concentrations remained the same between the two sampling periods or were higher during the stressed period as was the case for the Northern Kentucky and Johnson County water treatment plants. It is possible that during periods of lower organic carbon concentrations in the influent, there are fewer reaction or interaction sites for the metal salts resulting in elevated concentrations of these materials. However natural fluctuations of these materials in the raw water could also affect the residual concentrations. Concentrations of the other inorganic materials did not vary considerably between the stressed and unstressed periods, with the exception of the calcium and magnesium concentrations in some cases. The calcium concentration was higher during the unstressed period at the Houston plant while it was higher during the stressed period at the Northern Kentucky plant. The concentration of magnesium was elevated during the unstressed period at the Cincinnati plant. These fluctuations may have consequences with regards to headloss as the divalent cations may react with negatively charged organic materials to form larger complexes.

Table 5.6

Influent organic and inorganic concentrations measured at the various water treatment plants. Concentrations represent those found in a composite water sample and have not been fractionated according to size

	Milwaukee		**Modesto**		**Houston**		**Kentucky**	
	Stress	Unstress	Stress	Unstress	Stress	Unstress	Stress	Unstress
Ca (ppm)	36.7	35.4	3.91	4.02	47	54.2	34.6	28.3
Mg (ppm)	11.7	16.6	1.35	1.21	3.68	3.90	10.8	8.47
Al (ppm)	0.69	1.80	0.13	0.13	0.84	0.14	0.23	0.19
Fe (ppm)	-	0.52	0.12	-	0.07	-	0.75	0.40
Si (ppm)	1.43	1.22	6.8	6.60	6.3	6.65	6.13	3.75
Mn (ppm)	0.001	-	0.005	-	0.012	0.009	0.003	0.003
TOC (ppm)	2.39	1.82	1.47	1.64	4.88	6.86	1.8	1.8
	Cincinnati		**Johnson**		**Nevada**		**Buffalo**	
	Stress	Unstress	Stress	Unstress	Stress	Unstress	Stress	Unstress
Ca (ppm)	37.6	32.1	25.4	20.9	80.2	80.2	51.6	-
Mg (ppm)	11.6	23.7	16.8	14.4	28.0	28.7	23.7	-
Al (ppm)	0.25	0.81	0.34	0.10	0.05	-	0.81	-
Fe (ppm)	0.28	-	0.075	0.11	0.04	0.04	-	-
Si (ppm)	2.92	0.64	16	7.95	7.80	7.82	0.64	-
Mn (ppm)	0.02	0.045	-	-	0.001	0.001	0.045	-
TOC (ppm)	3.32	2.25	2.83	2.22	3.16	3.08	6.5	-

Filter Run-time

Filter run-time as a function of the filter mass loading rate of organic carbon during stressed and unstressed periods is reported in Figure 5.18. From the data shown here there is a general trend of decreasing filter run-time with increasing organic carbon mass loading rate. Furthermore, the data falls into two distinct groups, with both following the same overall trend of decreasing run-time with increasing mass loading rate. One group of utilities consisting of Milwaukee, Nevada, and Houston, during the stressed period, had run-times that were 100 hours or longer. The second group of utilities (Johnson County, Cincinnati, Northern Kentucky, Modesto, and Houston) had filter run-times that were less than 100 hours during the unstressed period.

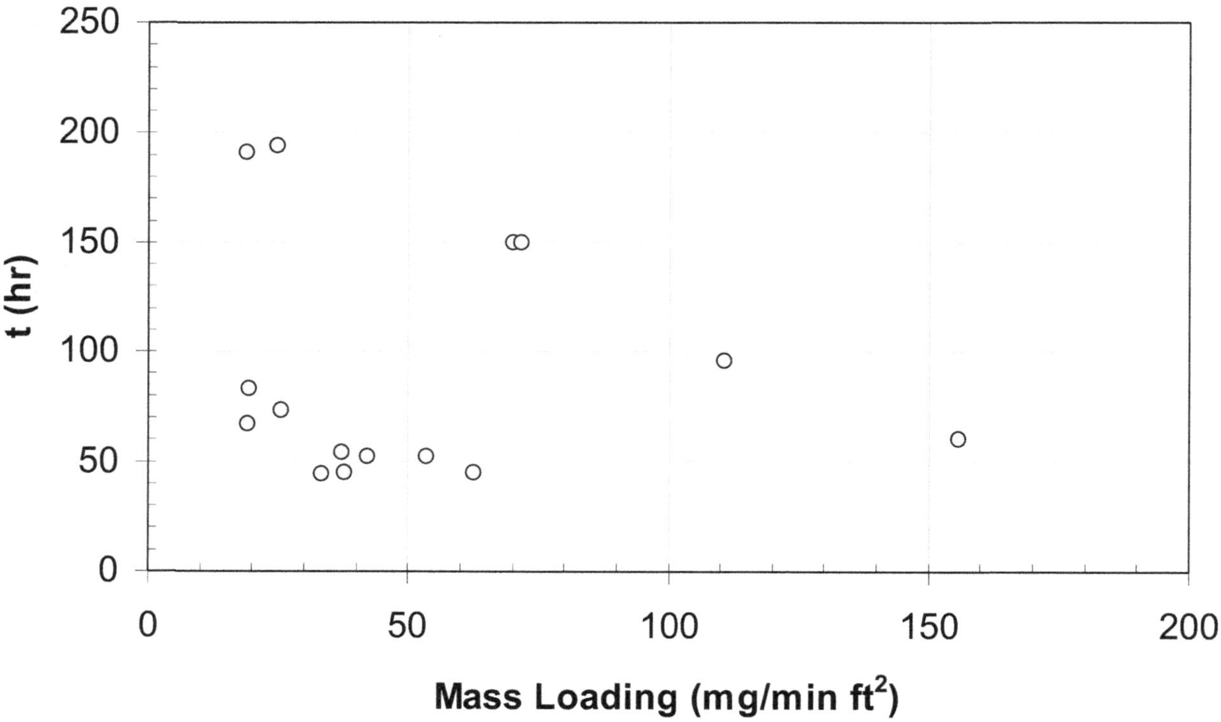

Figure 5.18 Filter run time plotted as a function of the mass loading rate of organic carbon during both stressed and unstressed operating periods for each of the participating water treatment plants

Effect of Filter Influent Ca/TOC Ratio on Filter Run-time

The ratio of concentrations of calcium and TOC in the filter influent (Ca/TOC) may have an effect on influent particle size fractionation due to the complexation between each other and may also affect the filter run times. Ca/TOC ratios and filter run times are given in Tables 5.7 and 5.8, respectively. Additionally, filter run time is plotted as a function of the ratio of respective calcium and TOC concentrations for the selected utilities in Figure 5.19. When the Ca/TOC ratio is above approximately 25, a significant increase in filter run time is evident. This appears to occur for both the stressed and unstressed periods. One possible reason could be that organic material not complexed at lower Ca/TOC ratios may make the filter media surface sticky and this may decrease the filter run-times.

Table 5.7
Ratio of calcium to organic carbon concentrations in the filter influent at selected water treatment plants. Concentrations represent those found in a composite water sample and have not been fractionated according to size

	Milwaukee		**Modesto**		**Houston**		**Kentucky**	
	Stress	Unstress	Stress	Unstress	Stress	Unstress	Stress	Unstress
Ca (ppm)	36.7	35.4	3.91	4.02	47	54.2	34.6	28.3
Ca (mg $CaCO_3$/l)	91.75	88.5	9.775	10.05	117.5	133	86.5	70.75
TOC (ppm)	2.39	1.82	1.47	1.64	4.88	6.86	1.77	1.8
Ca/TOC (mg $CaCO_3$/mg C)	38.39	48.63	6.65	6.13	24.08	19.75	48.87	39.31
	Cincinnati		**Johnson**		**Nevada**		**Buffalo**	
	Stress	Unstress	Stress	Unstress	Stress	Unstress	Stress	Unstress
Ca (ppm)	37.6	32.1	25.4	20.88	80.2	80.2	51.6	-
Ca (mg $CaCO_3$/l)	94	80.25	63.5	52.2	200.5	200.5	129	-
TOC (ppm)	3.32	2.25	2.83	2.22	3.16	3.08	6.5	-
Ca/TOC (mg $CaCO_3$ / mg C)	28.31	35.67	22.44	23.51	63.45	65.01	19.85	-

Table 5.8
Filter run times for selected treatment facilities. The reported values represent the long-term historical averages for these facilities

	Milwaukee		**Modesto**		**Houston**		**Kentucky**	
	Stress	Unstress	Stress	Unstress	Stress	Unstress	Stressed	Unstress
t (hour)	194	191	44	54	96	60	67	83
	Cincinnati		**Johnson**		**Nevada**		**Buffalo**	
	Stress	Unstress	Stress	Unstress	Stress	Unstress	Stressed	Unstress
t (hour)	45	73	52	52	150	150	45	-

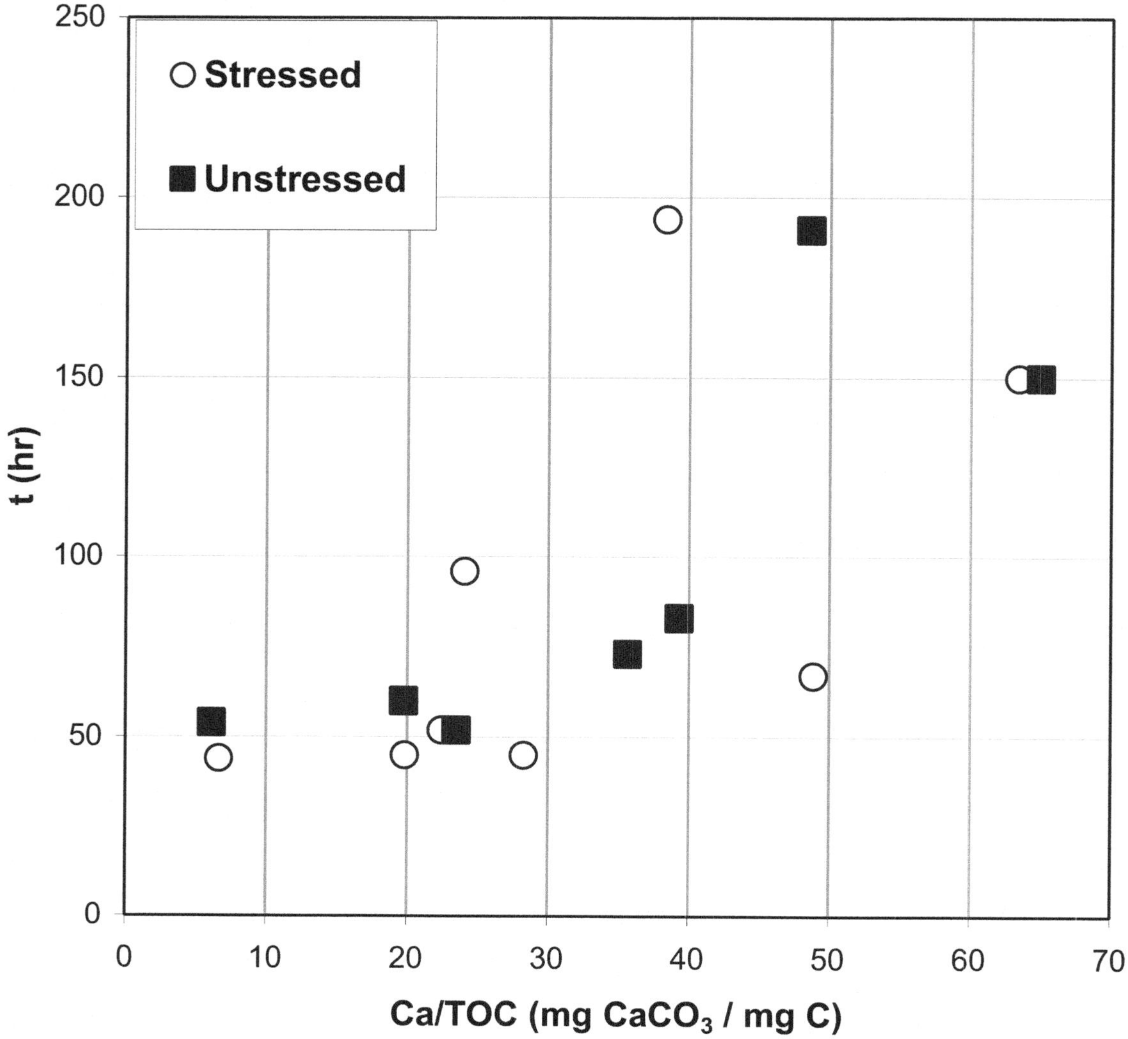

Figure 5.19 Filter run time as a function of the calcium to TOC ratio in the filter influent. The calcium and TOC concentrations are the total concentration of each in a composite influent water sample

Rapid and Slow Filtration

The Salem water treatment plant is the only utility in this study that utilizes slow sand filtration. Under certain circumstances slow sand filtration has a number of advantages over rapid filtration, specifically its simplicity of operation (Crittenden, et al. 2005, Gorczyca and London 2003). The organic and inorganic particle size fractions for the filter influent and effluent for this slow sand facility are reported in Figure 5.20 and 5.21, respectively, for the stressed period. Organics in the filter influent are present in substantial quantities in each of the four size fractions, with the dissolved + fraction being the largest followed by the filtrable, ultrafine, and settleable fractions, respectively. For the filter effluent, the dissolved + organic size fraction was largest followed by the ultrafine, settleable, and filtrable size fractions, respectively.

The concentration of organic material in the filter influent and effluent essentially remained unchanged at 0.66. for the Salem water treatment plant. The percentage of effluent organic material in the dissolved + and filtrable size classes decreased, while that in the settleable and ultrafine size classes increased. Most of the inorganic material in the filter influent and effluent was present in the dissolved + size class (Figure 5.21).

Inorganic concentrations in the filter influent and effluent are given in Figure 5.22 - 5.25, respectively. Iron and aluminum were predominantly present in the settleable, filtrable and ultrafine size classes, and were completely removed from the filter effluent (Figure 5.22). Silica was predominantly present in the dissolved + size class and was not removed by the slow sand filters (Figure 5.23). Manganese was primarily present in the settleable and dissolved + size fractions and approximately 75% was removed through filtration, with the remaining manganese in the effluent present in the settleable size class. Calcium and magnesium were present in the dissolved + size class in the influent and a slight increase in the concentrations of these ions was observed in the effluent (Figure 5.24). Arsenic and copper, present in the settleable and filtrable size classes, were completely removed.

The relative size distribution of organic and inorganic material in the filter influent and effluent for Salem's slow sand filters and average values for all the other utilities that use rapid sand filters is shown in Figures 5.26 and 5.27. The organic material in the rapid sand filters influent and effluent is predominantly present in the dissolved + size fraction. For the slow sand filter, the influent organic material is predominantly present in the dissolved +, ultrafine and filtrable size fractions, while the organic material in the effluent was present in the dissolved +, ultrafine and settleable size classes (Figure 5.26). While the organic fraction of dissolved + materials was about 67% and 73% for influent and effluent respectively for rapid sand filtration, it was about 38% and 32% in the influent and effluent, respectively, for slow sand filtration. The presence of organic material in the larger size classes in the slow sand filter effluent may be due to the detachment of biofilms occurring in the slow sand filter media. Inorganic material was primarily (80%) present in the dissolved + size fraction in the filter influent and effluent for those facilities using rapid sand filtration. Most of the inorganic material in the slow sand filter's influent and effluent was also present in the dissolved + size fraction. However, in the slow sand filter effluent, almost all the inorganic material (~96%) was present in the dissolved + size fraction (Figure 5.27).

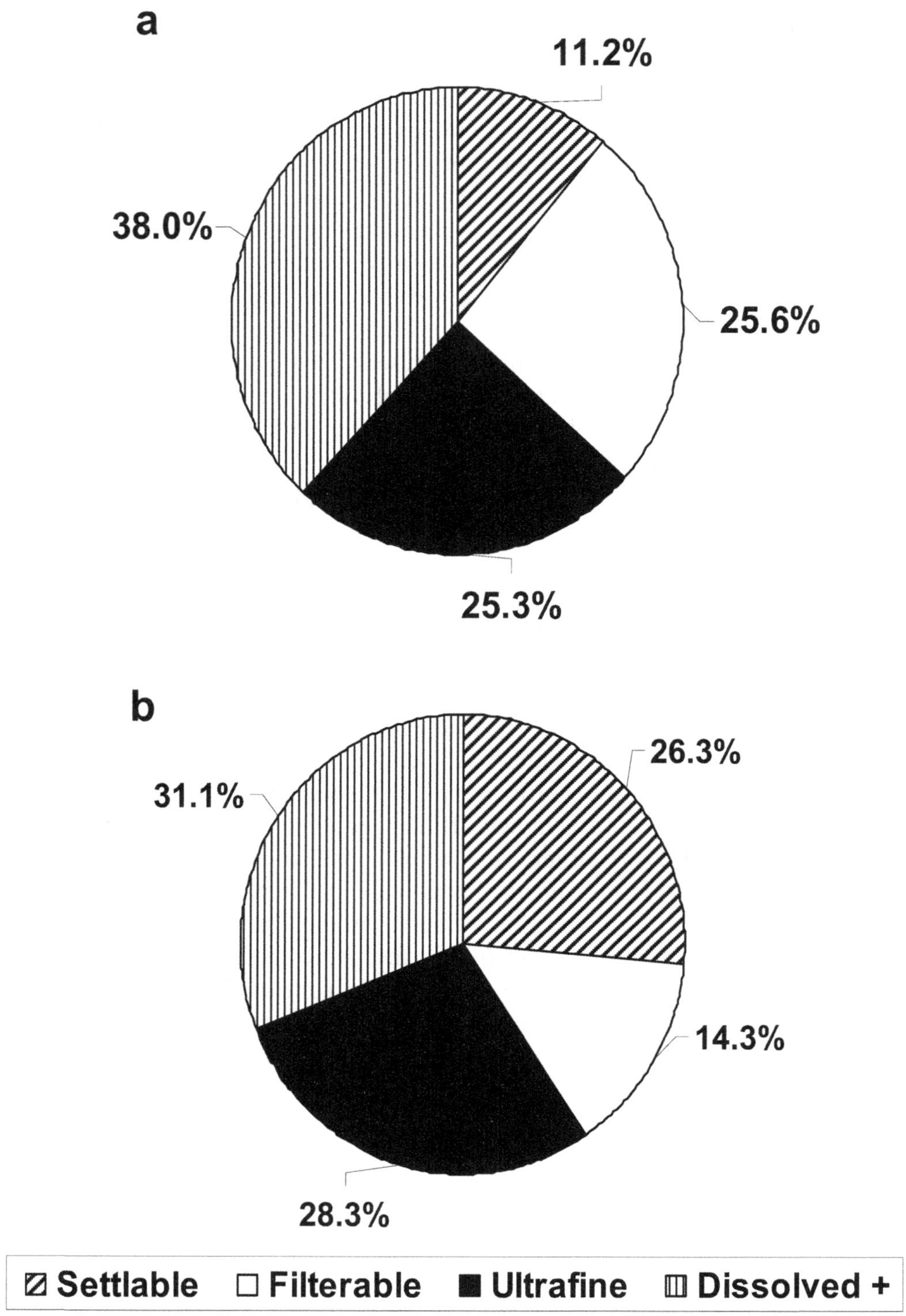

Figure 5.20 Relative size distributions of organics in the filter (a) influent and (b) effluent at the Salem water treatment plant during the stressed operating period (influent $[TOC]_T$ = 0.673 ppm, effluent $[TOC]_T$ = 0.663 ppm)

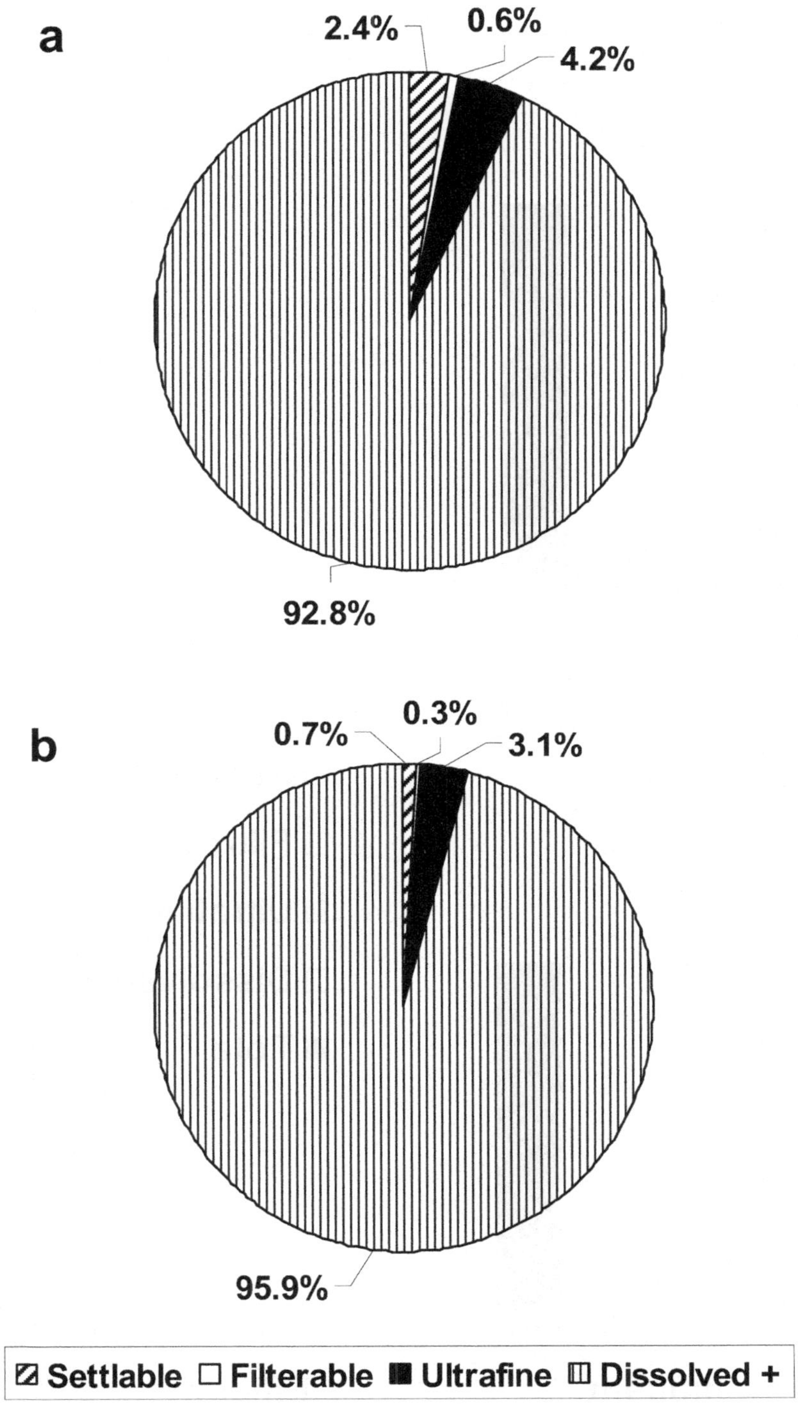

Figure 5.21 Relative size distributions of inorganics in the filter (a) influent and (b) effluent at the Salem water treatment plant during the stressed operating period (influent $[Metals]_T$ = 15.71 ppm, effluent $[Metals]_T$ = 15.50 ppm)

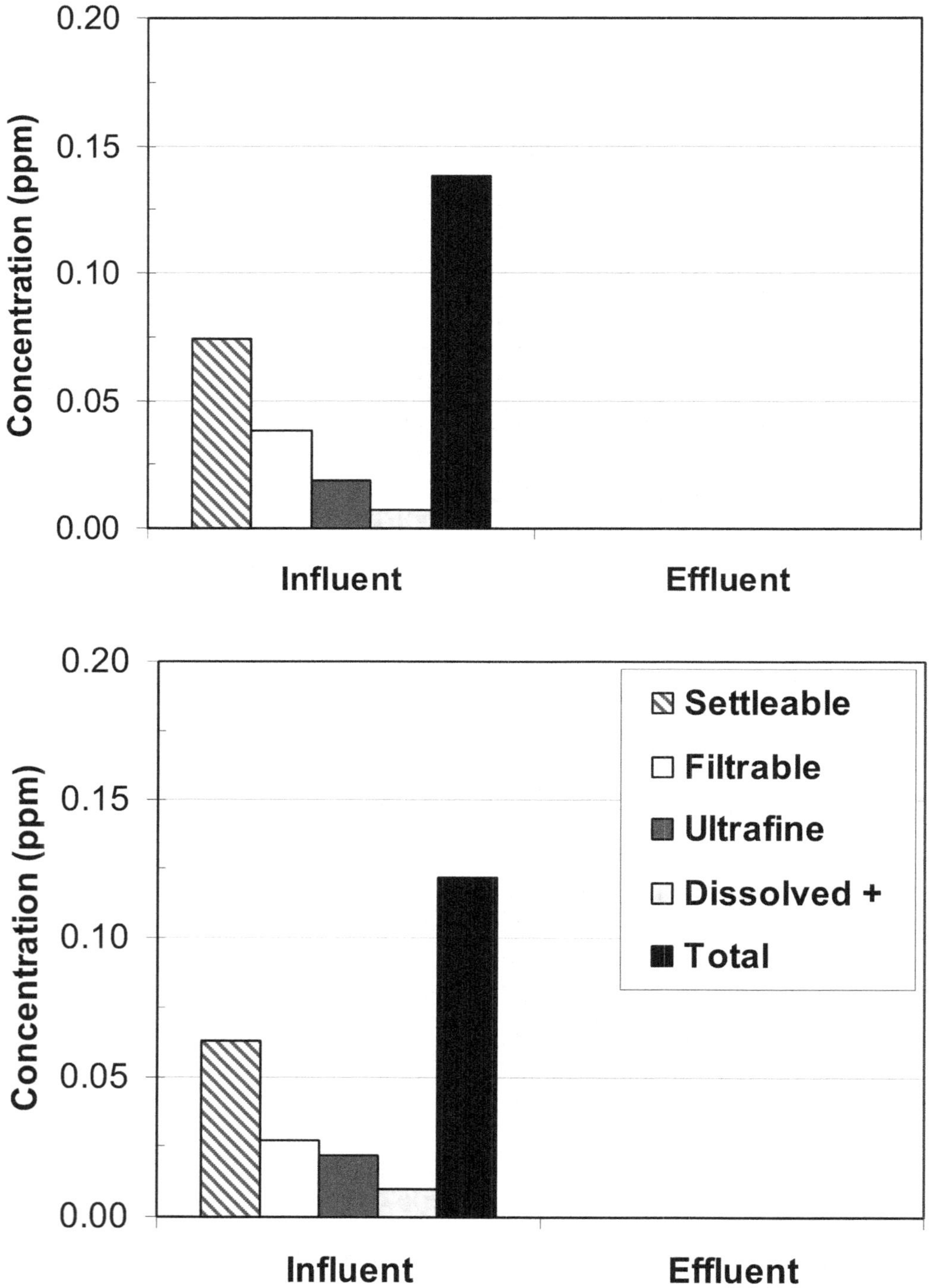

Figure 5.22 Filter influent and effluent concentrations in the different size fractions for (a) iron and (b) aluminum at the Salem water treatment plant

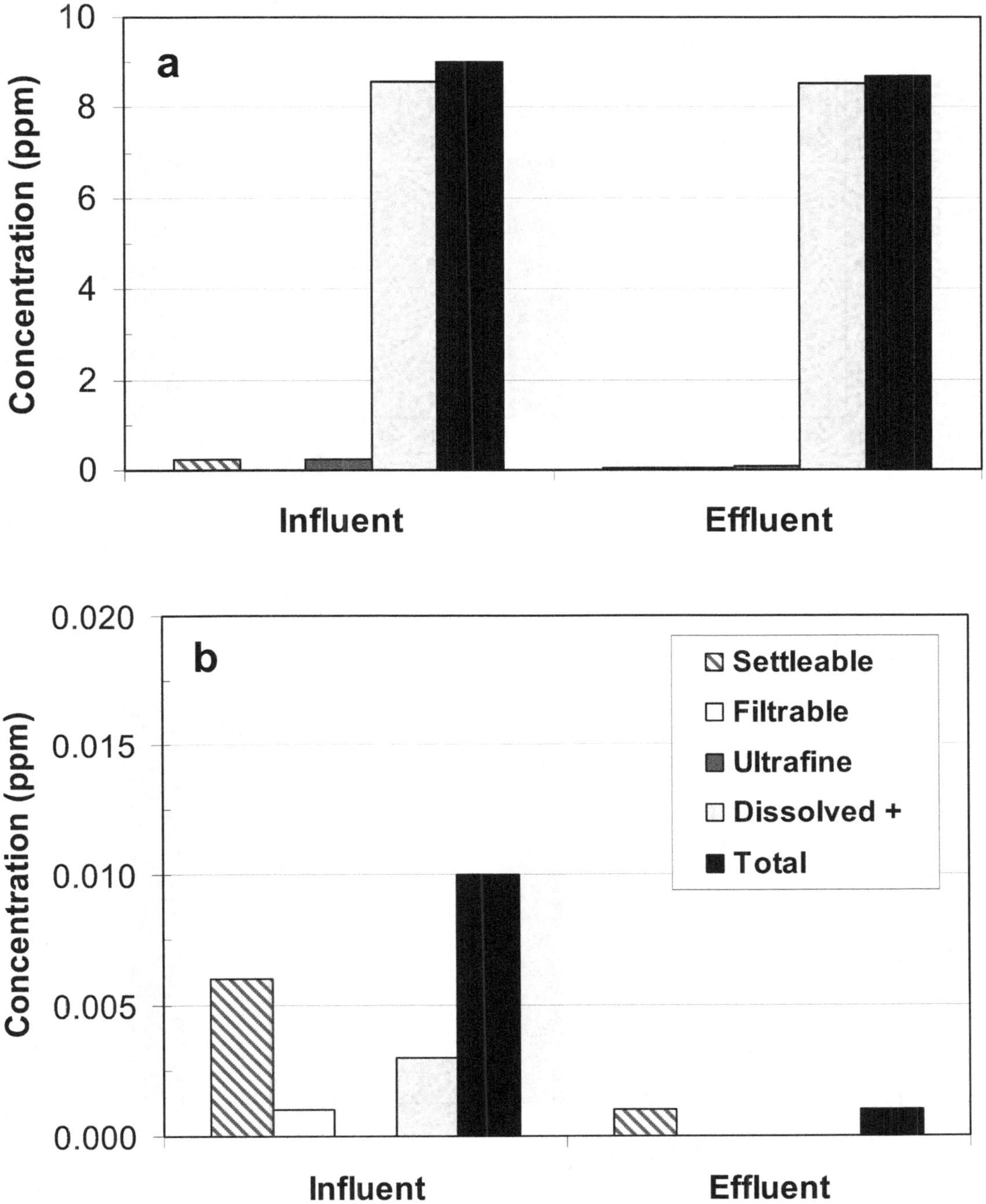

Figure 5.23 Filter influent and effluent concentrations in the different size fractions for (a) silica and (b) manganese at the Salem water treatment plant

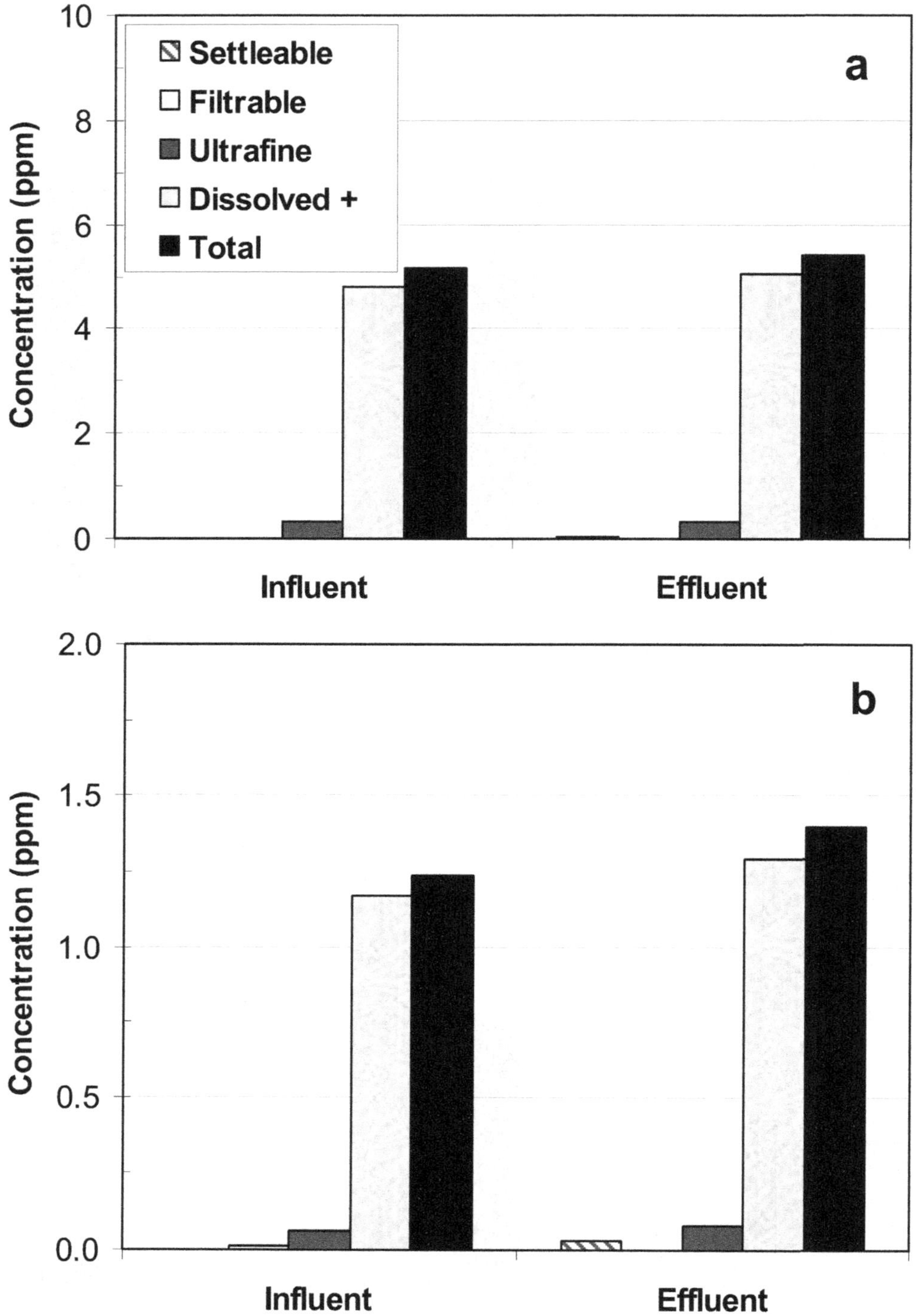

Figure 5.24 Filter influent and effluent concentrations in the different size fractions for (a) calcium and (b) magnesium at the Salem water treatment plant

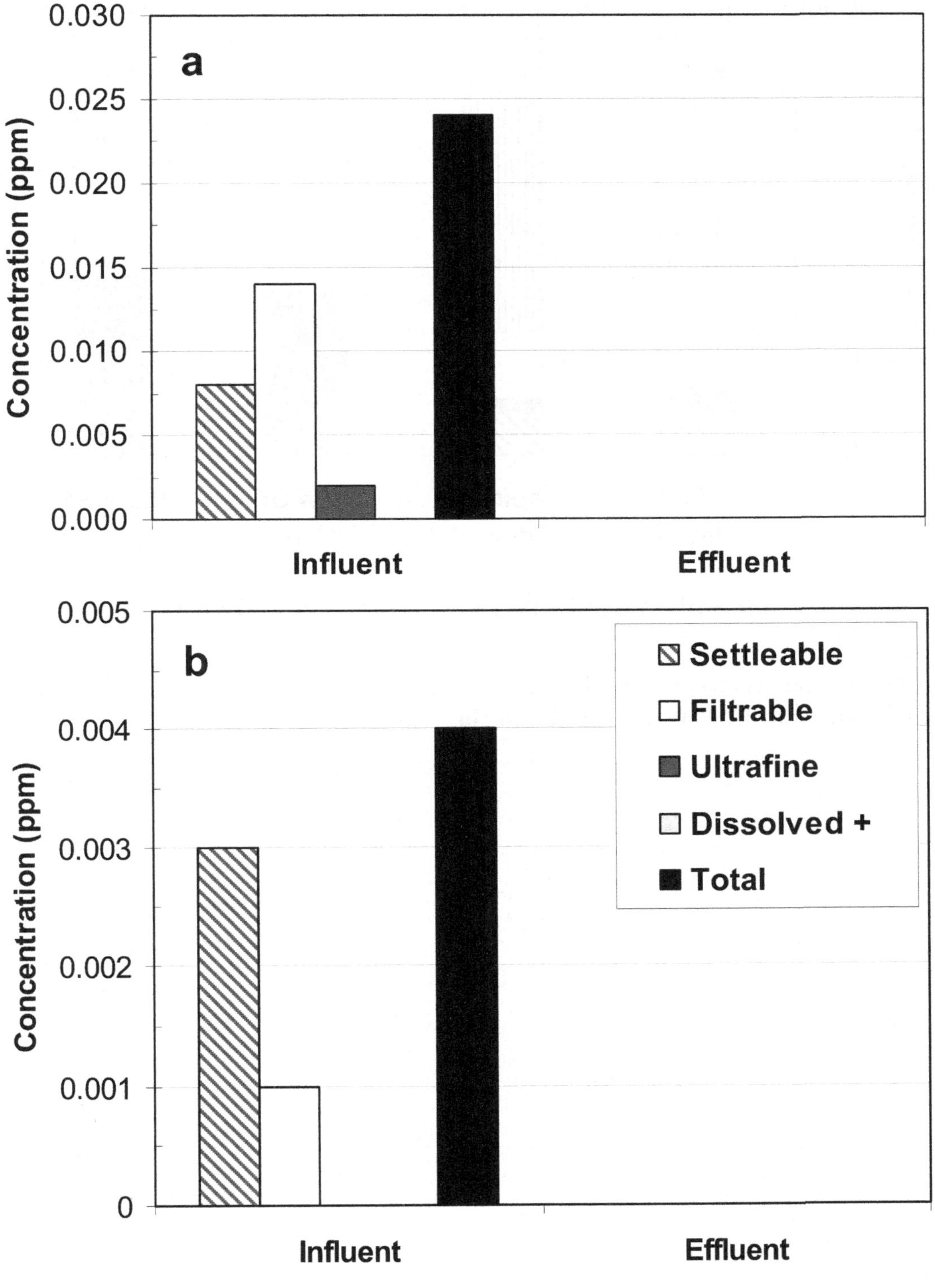

Figure 5.25 Filter influent and effluent concentrations in the different size fractions for (a) arsenic and (b) copper at the Salem water treatment plant

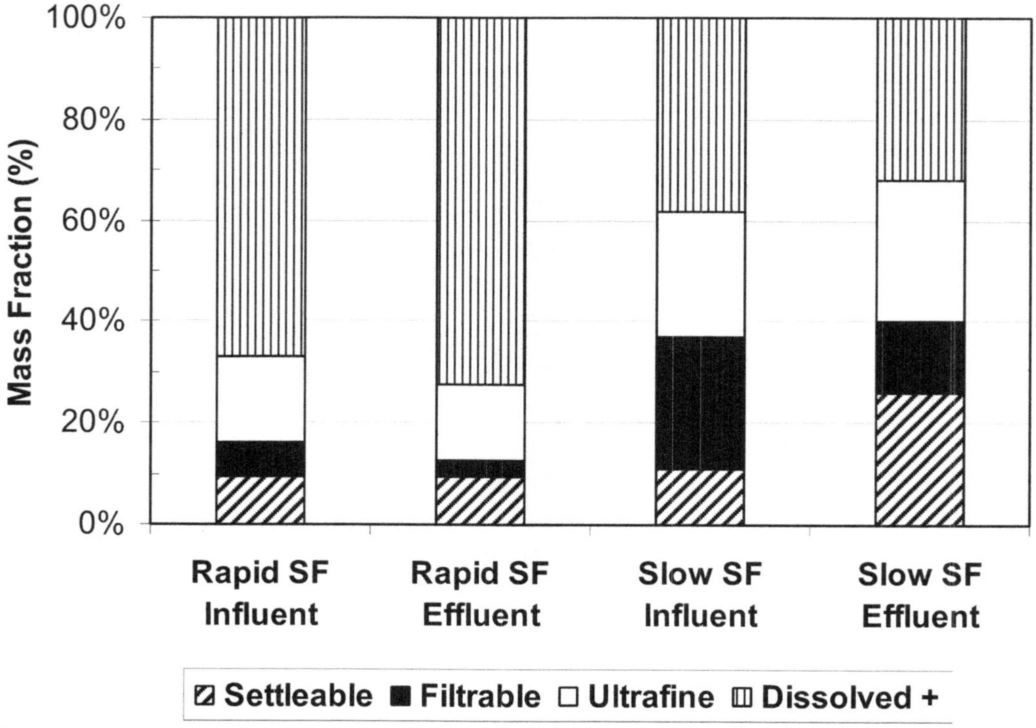

Figure 5.26 Mean relative size distribution of organic material in the filter influent and effluent for those utilities using rapid and slow sand filtration

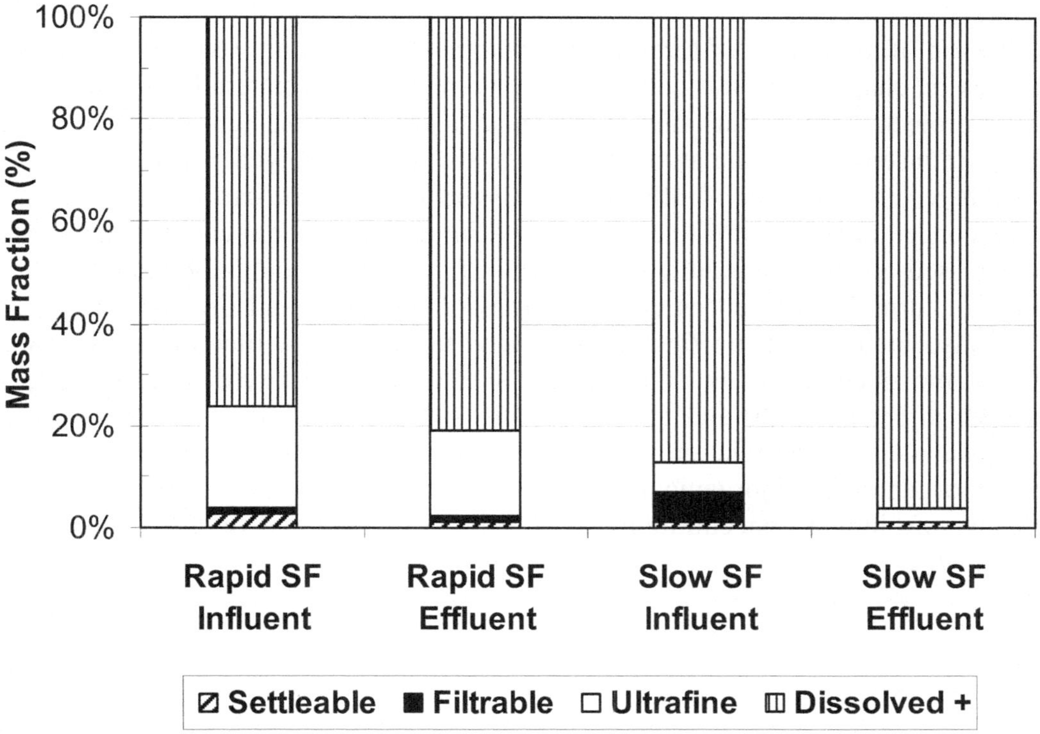

Figure 5.27 Mean relative size distribution of inorganic material in the filter influent and effluent for those utilities using rapid and slow sand filtration

Relationship between Turbidity and Particle Size Distribution

An evaluation of the ability of turbidity measurements commonly recorded at the water treatment facilities to characterize the concentration or removal of materials was one of the objectives of this project. Because turbidity measurements are based on light blockage we may assume that they are insensitive to the amount of materials in the dissolved + size range and are thus not included here. The turbidity of influent and effluent for all of the participating utilities as a function of the concentration of ultrafine organic and inorganic material concentrations under unstressed and stressed operating conditions is shown in Figures 5.28 and 5.29, respectively. To extend the range of turbidity as well as concentrations of inorganic and organic material, the influent data are also included in the figures. Similarly, turbidity as a function of the sum of inorganic and organic materials in the three larger size classes (i.e., settleable, filtrable and ultrafine) is shown in Figures 5.30 and 5.31 for unstressed and stressed operating conditions, respectively. As is evident, no correlation was observed between the measured turbidity and the concentrations of organic and inorganic materials in various size classes. Indeed, in some instances, water having higher concentrations of ultrafine inorganic and organic materials had the lower finished water turbidity.

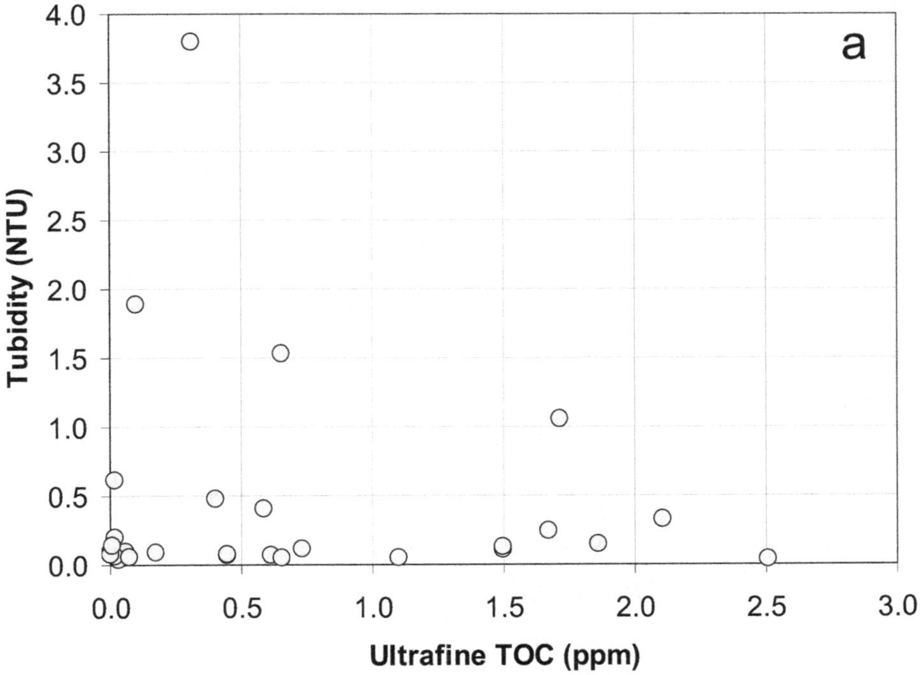

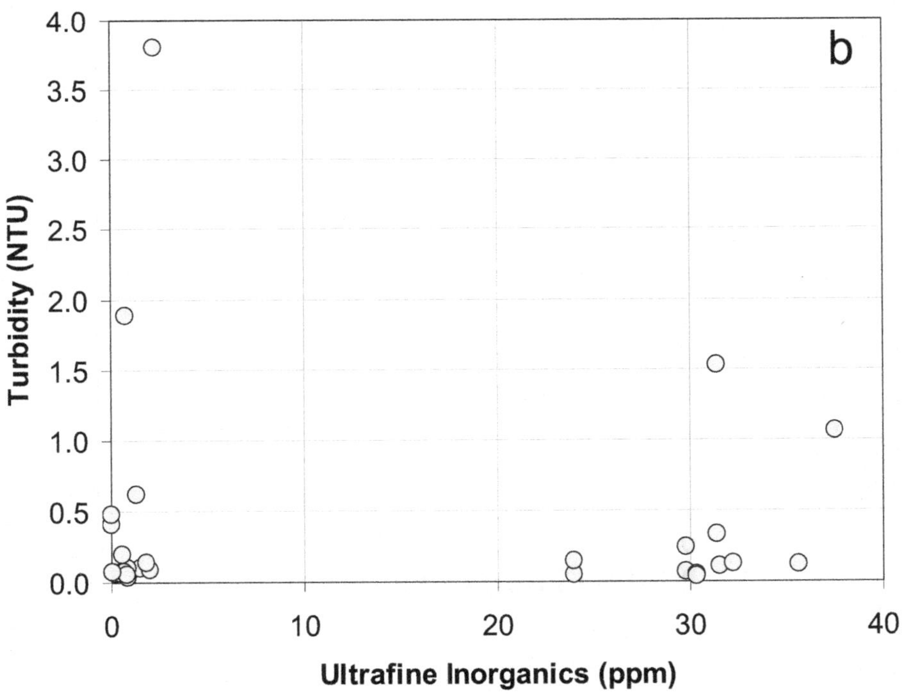

Figure 5.28 Influent and effluent turbidity for all utilities as a function of the concentration of (a) ultrafine organic and (b) ultrafine inorganic materials in the filter influent and effluent during the unstressed operating period

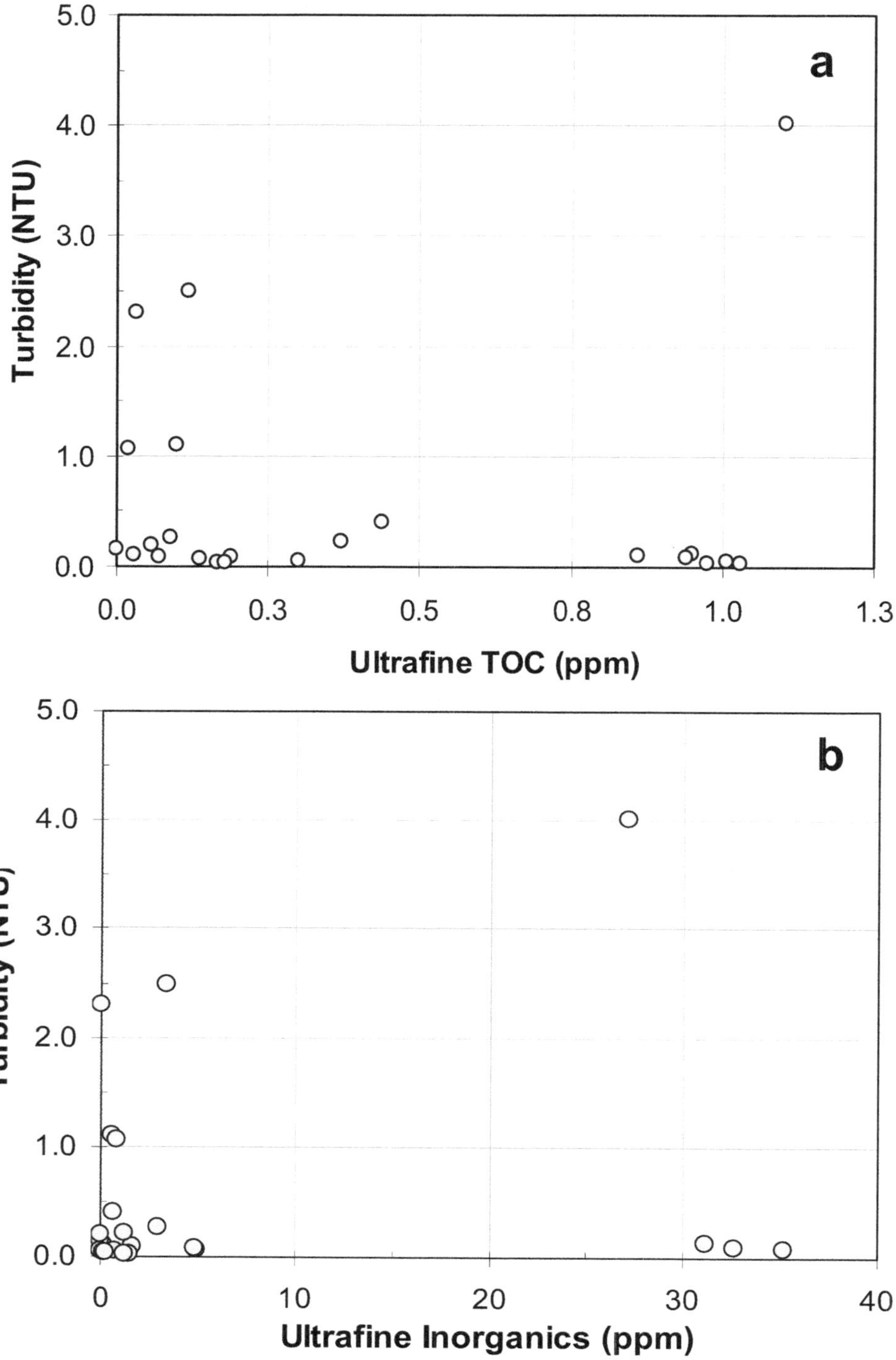

Figure 5.29 Influent and effluent turbidity for all utilities as a function of the concentration of (a) ultrafine organic and (b) ultrafine inorganic materials in the filter influent and effluent during the stressed operating period

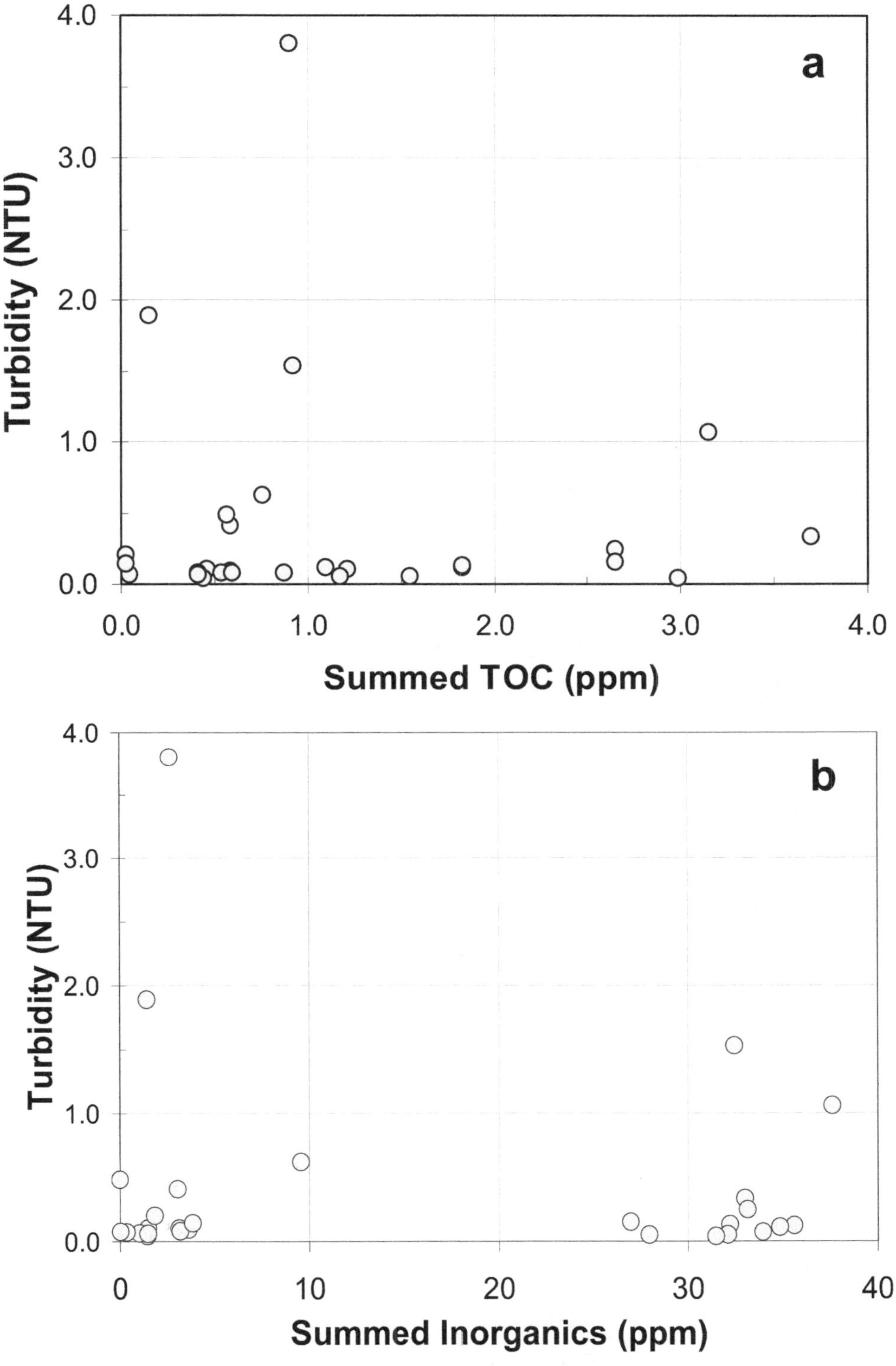

Figure 5.30 Influent and effluent turbidity for all utilities as a function of the sum concentration of (a) organic and (b) inorganic materials in the settleable, filtrable and ultrafine size classes in the filter influent and effluent during the unstressed operating period

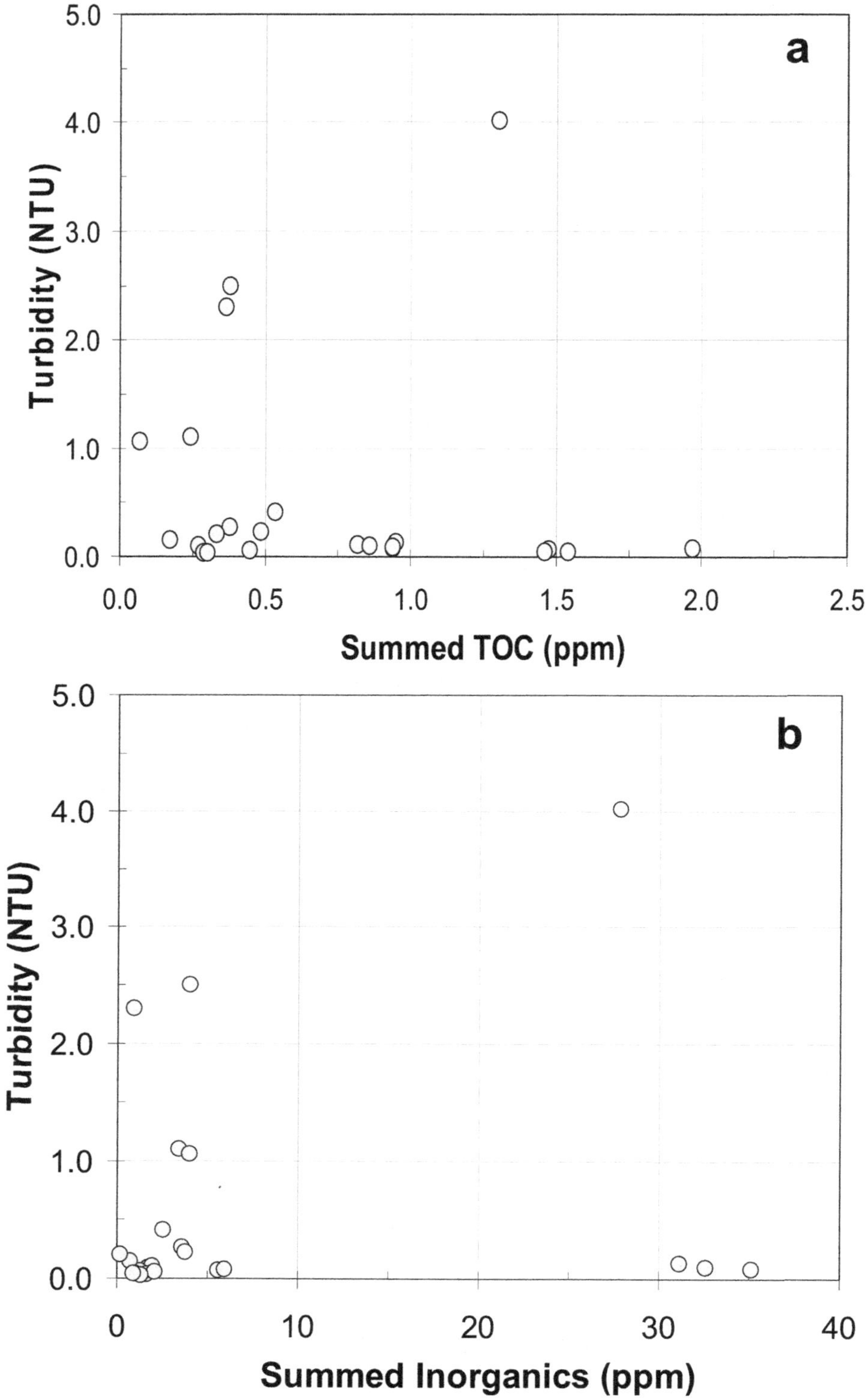

Figure 5.31 Influent and effluent turbidity for all utilities as a function of the sum concentration of (a) organic and (b) inorganic materials in the settleable, filtrable and ultrafine size classes in the filter influent and effluent during the stressed operating period

Ultrafine Size Fraction

Relationship between Organic and Inorganic Ultrafine Size Fractions

To investigate correlation between organic and inorganic material in the ultrafine size class, the concentrations of ultrafine organic and inorganic material are plotted against one another in Figure 5.32 for the filter influent and effluent for all of the participating utilities. For the filter effluent, data are shown for various stages of filter run (Figures 5.32b-c). Because of low concentrations of organic and inorganic material in this size class for several samples, most of the data are cluttered near the origin. Even though some correlation is evident using least squares fitting, it is evident that the few outliers dominate the correlation. Nevertheless, it appears that the larger amount of ultrafine inorganic material may be associated with higher organic mass in the same size class. It should be noted that at one facility, Northern Kentucky, the influent did not have any organic material in the ultrafine size class, but substantial amount was present in the effluent. The ultrafine inorganic material in the effluent of this particular utility is relatively low compared to the ultrafine organic material (the data points close to the X-axis near organic material concentrations of 0.3 to 0.5 mg/L are those observed at Northern Kentucky). The ultrafine organic and inorganic mass fractions are reported in Figure 5.33 for both the filter influent and effluent combined. A nearly linear relationship between the percent of ultrafine organic and inorganic material is observed. Further analyses such as actual interaction of organic material with specific inorganic constituents that contributed to such a relationship is required, and is beyond the scope of this study.

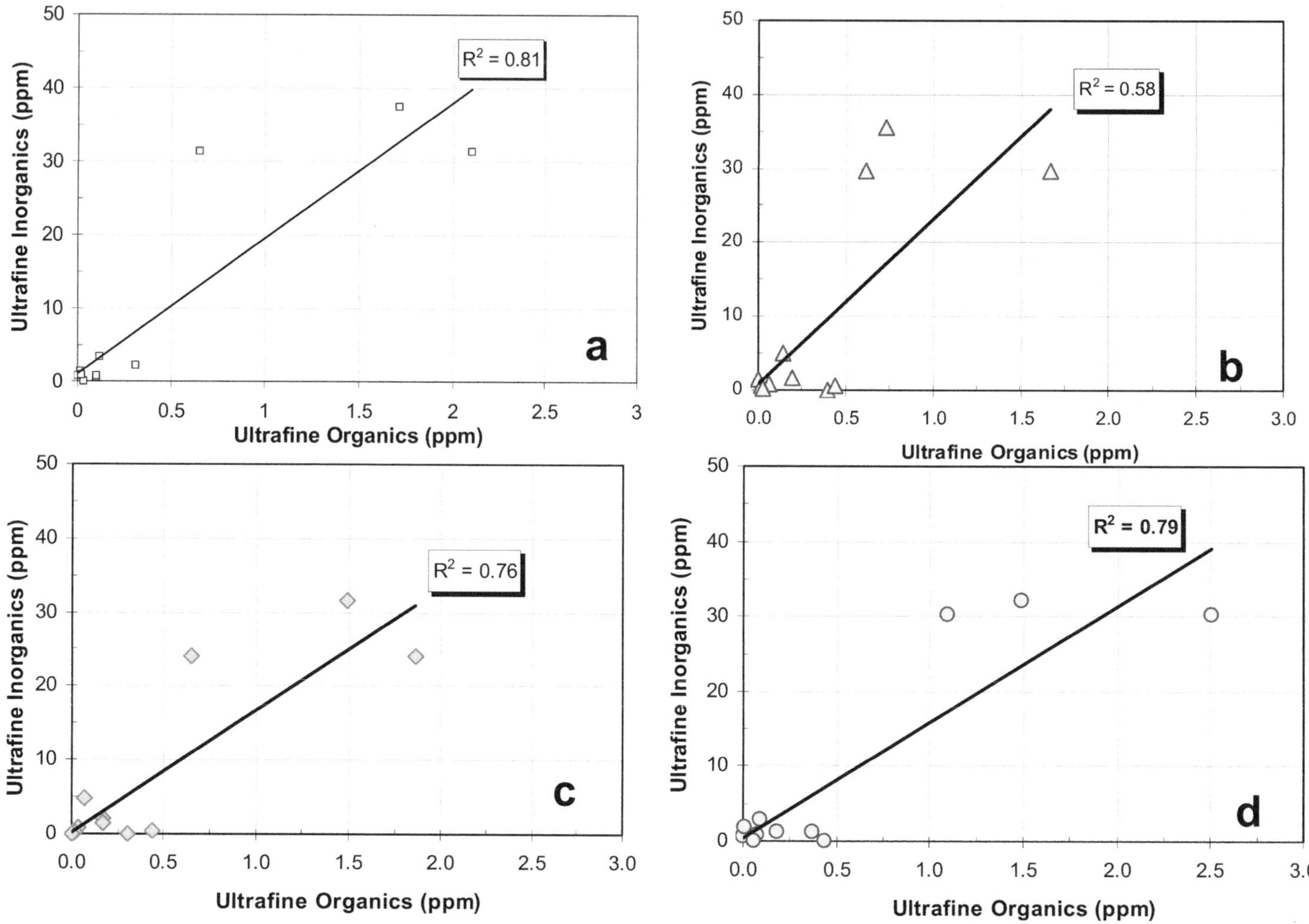

Figure 5.32 Concentration of ultrafine inorganic materials as a function of ultrafine organic materials for all utilities in the filter (a) influent and in the effluent at the (b) start, (c) middle, and (d) end of the filter runs

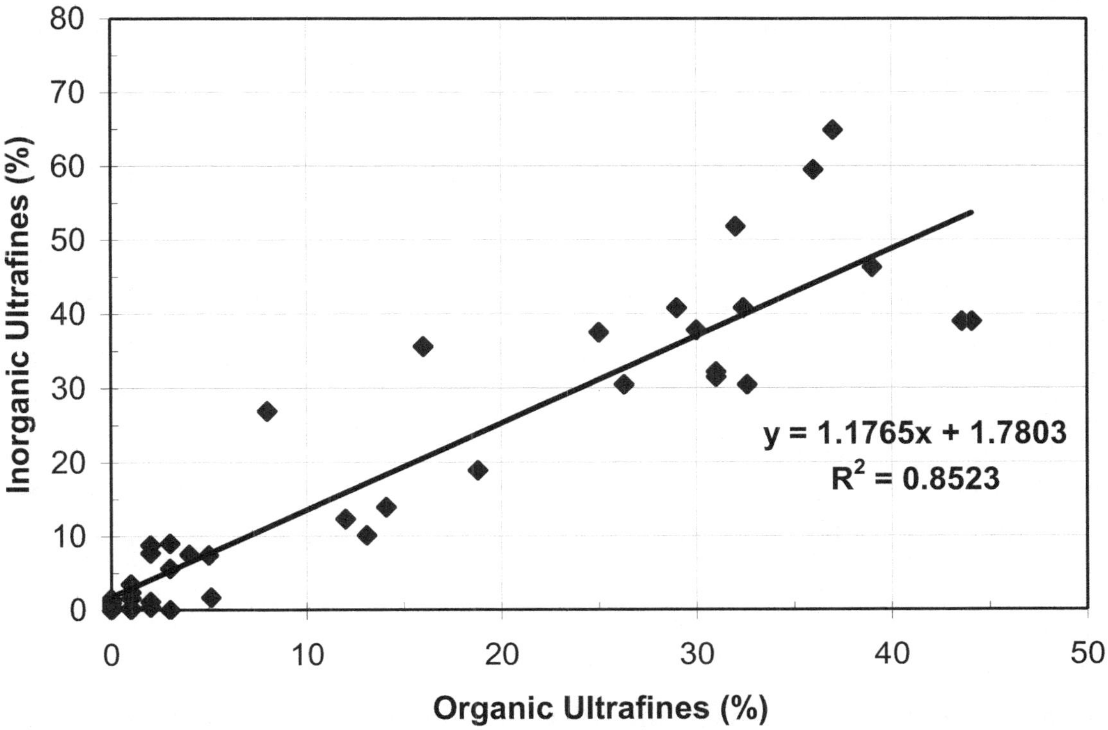

Figure 5.33 Mass fraction of inorganic ultrafines as a function of the mass fraction of organic ultrafines in the filter influent and effluent for each of the participating utilities.

Ultrafine Loading Rate and Filter Run Time

The loading rates of ultrafine inorganic and organic materials are plotted as a function of filter run time for all of the utilities during both stressed and unstressed conditions in Figure 5.34. It is evident that no direct correlation exists between the loading rates of either the ultrafine organic and inorganic materials and filter run-time. However, it appears that in general, filter run times tend to be lower at low ultrafine concentrations. The lack of a clear trend may be due to a number of factors including the relatively low concentration of ultrafine materials found in the different filter influents as discussed in Chapter 4.

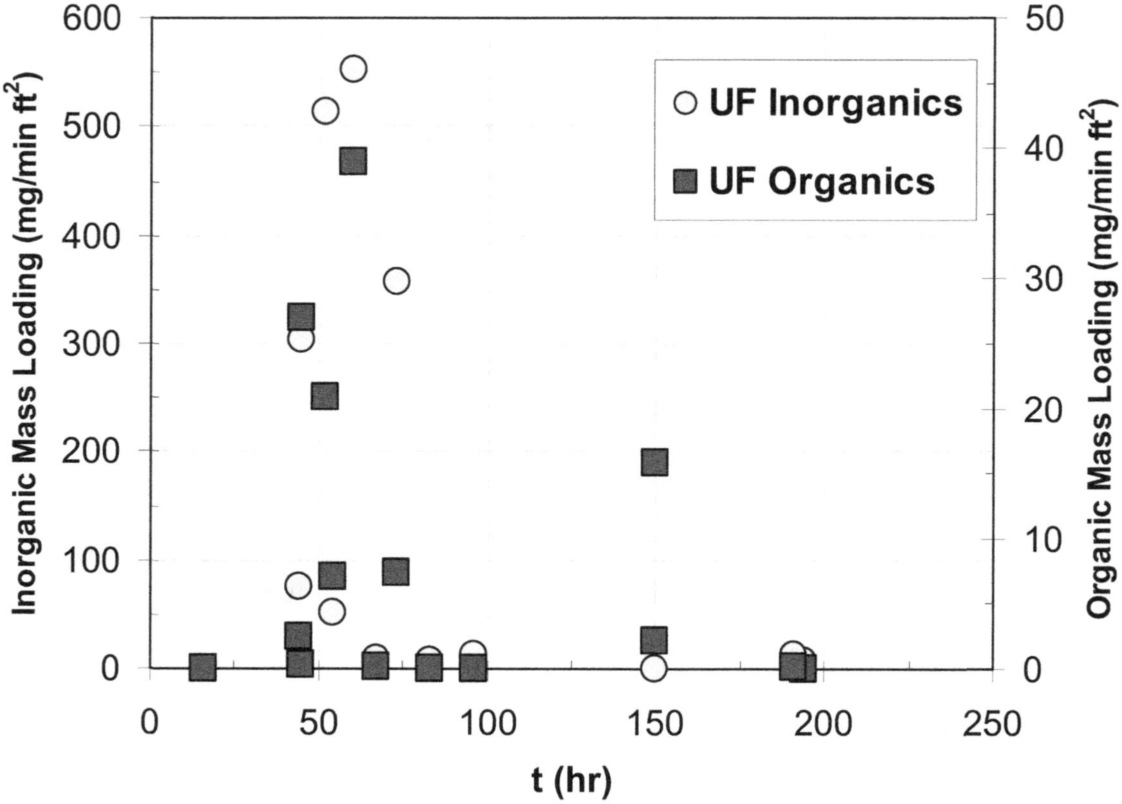

Figure 5.34 Relationships between the filter run time and the concentration of ultrafine organic and inorganic materials for all of the utilities. Data is reported for both the stressed and unstressed operating periods

Removal of Ultrafine Organics and Inorganics

The influent and effluent concentrations of TOC in the ultrafine fraction are given in Figures 5.35 - 5.42 for all plants during the stressed and unstressed periods. It was observed that in most of the cases, the TOC concentration in the ultrafine size fraction increased in the filter effluent during stressed period, while during unstressed periods, it decreased. There was a significant decrease in the ultrafine TOC concentration at the Modesto and Southern Nevada plants during unstressed periods. The degree to which the ultrafine TOC increased varied for the different plants and run times of the filters.

The concentrations of the inorganic material in the ultrafine size class in the filter influent and effluent during various stages of the filter runa are shown in Figure 5.43 - 5.50. Except for Southern Nevada, Houston, and Johnson County, the concentration of inorganic material in the ultrafine size class decreased during both stressed and unstressed operating conditions. In some instances, the ultrafine concentration increased during the filter run suggesting some elusion of this material. At the Johnson County and Southern Nevada plants, the concentration of the inorganic ultrafine fraction was higher in the effluent than the influent..

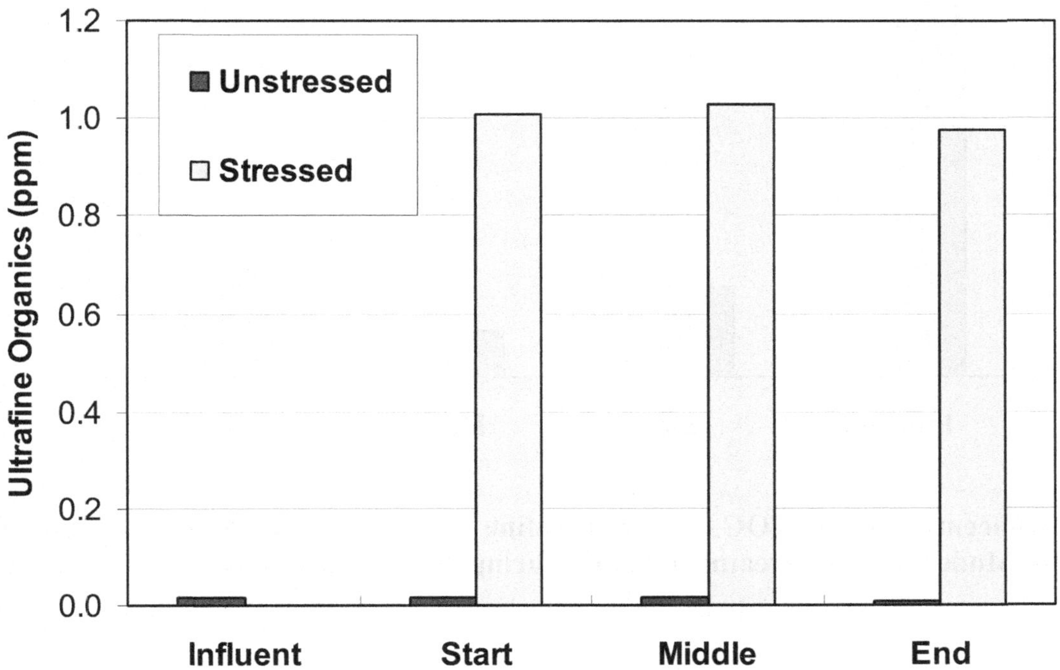

Figure 5.35 Size of the ultrafine fraction of TOC in the filter influent and effluent at the Milwaukee water treatment plant during stressed and unstressed operating conditions

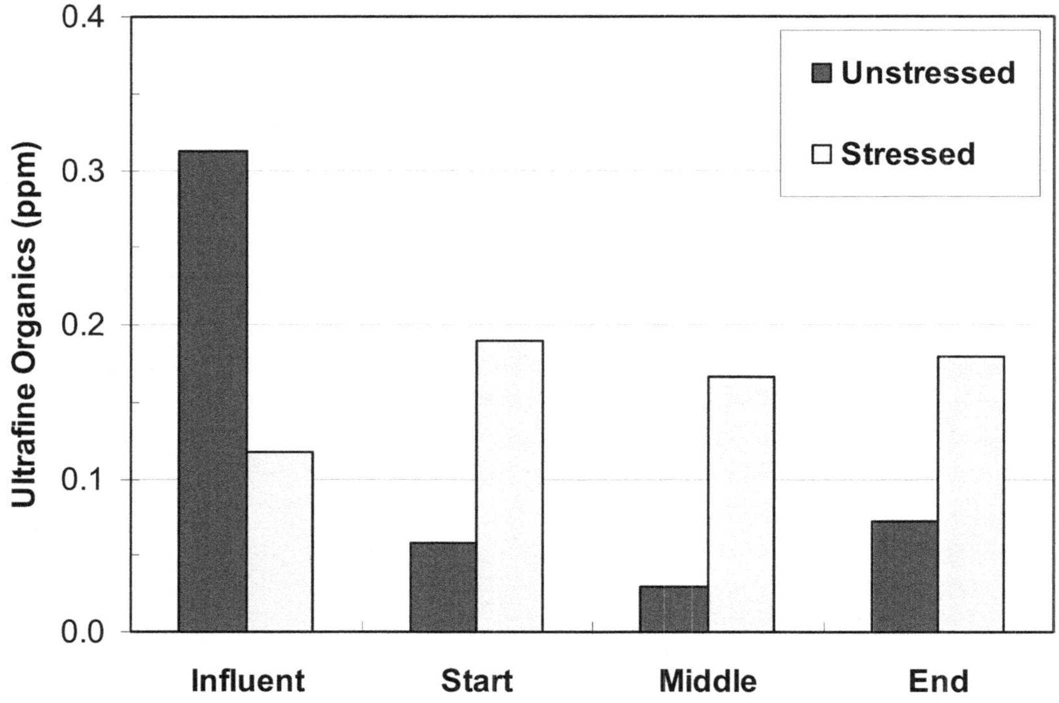

Figure 5.36 Concentration of TOC in the ultrafine fraction in the filter influent and effluent at the Modesto water treatment plant during stressed and unstressed operating conditions

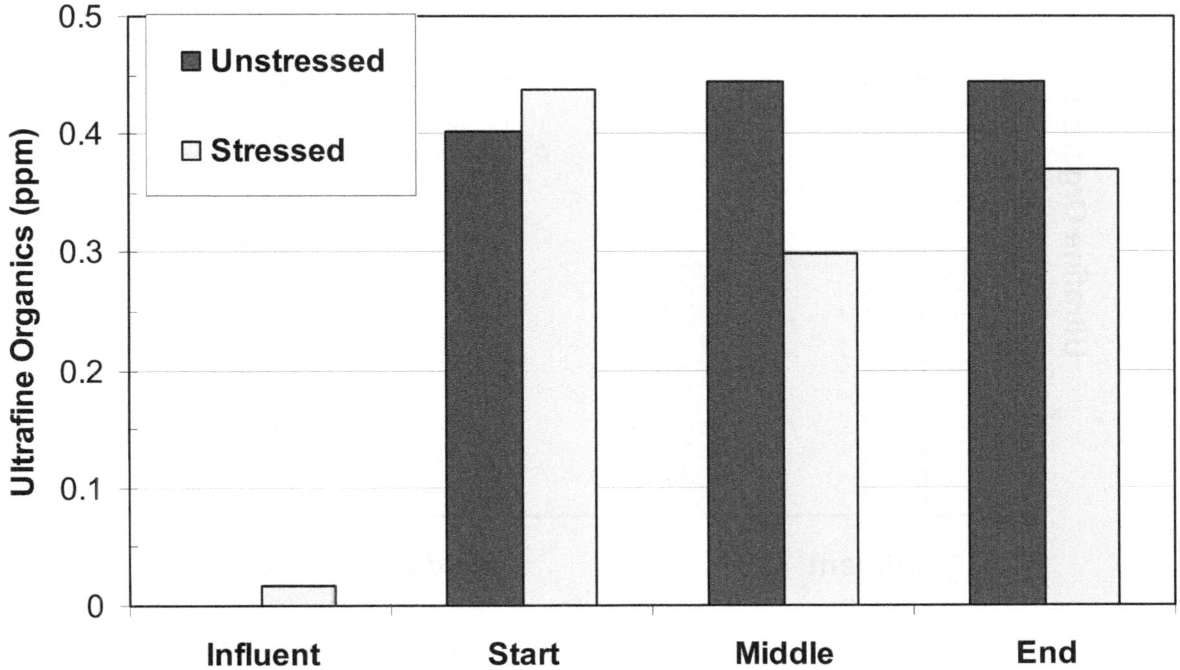

Figure 5.37 Concentration of TOC in the ultrafine fraction in the filter influent and effluent at the Northern Kentucky water treatment plant during stressed and unstressed operating conditions

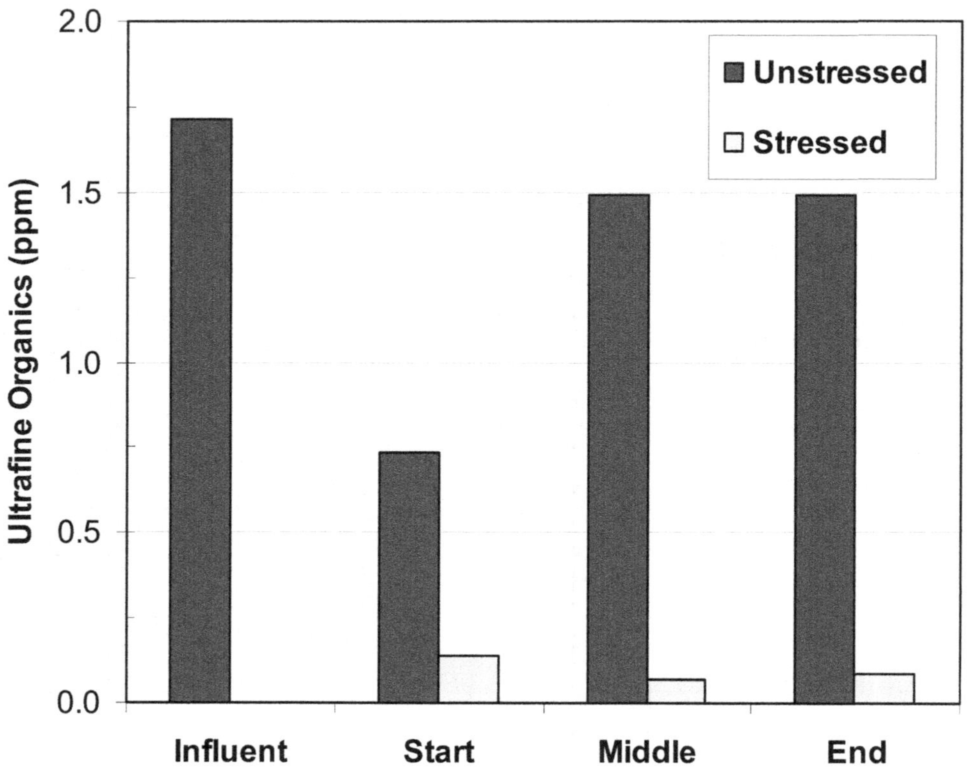

Figure 5.38 Concentration of TOC in the ultrafine fraction in the filter influent and effluent at the Houston water treatment plant during stressed and unstressed operating conditions

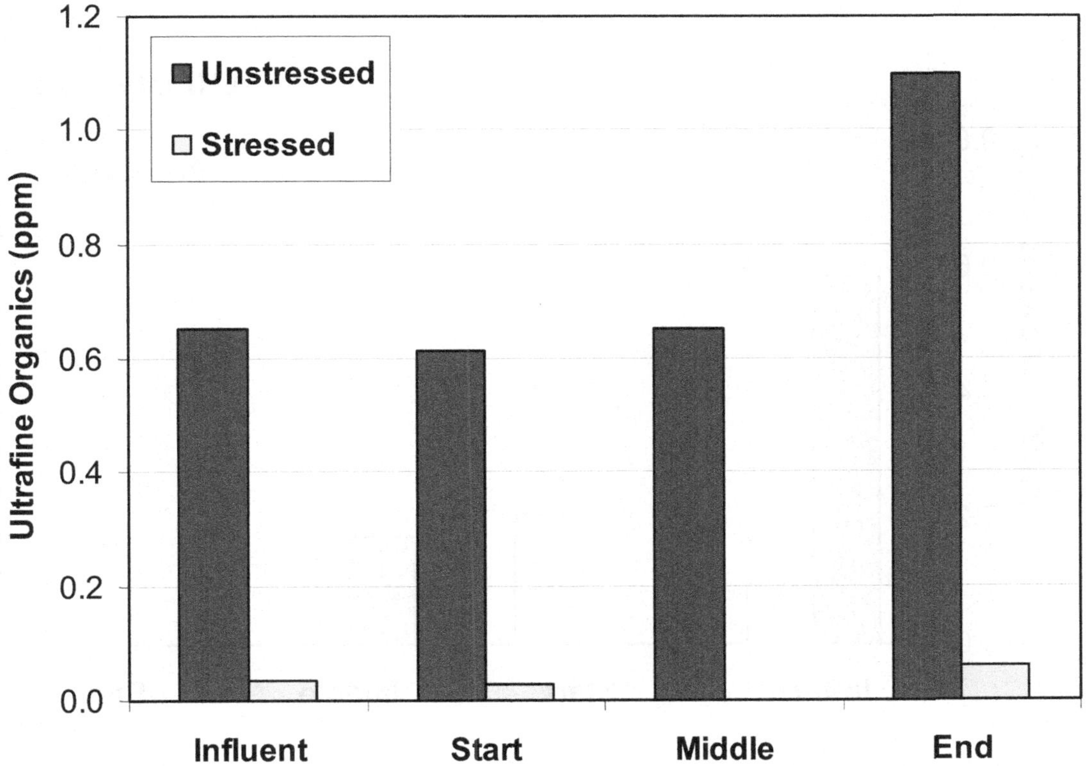

Figure 5.39 Concentration of TOC in the ultrafine fraction in the filter influent and effluent at the Cincinnati water treatment plant during stressed and unstressed operating conditions

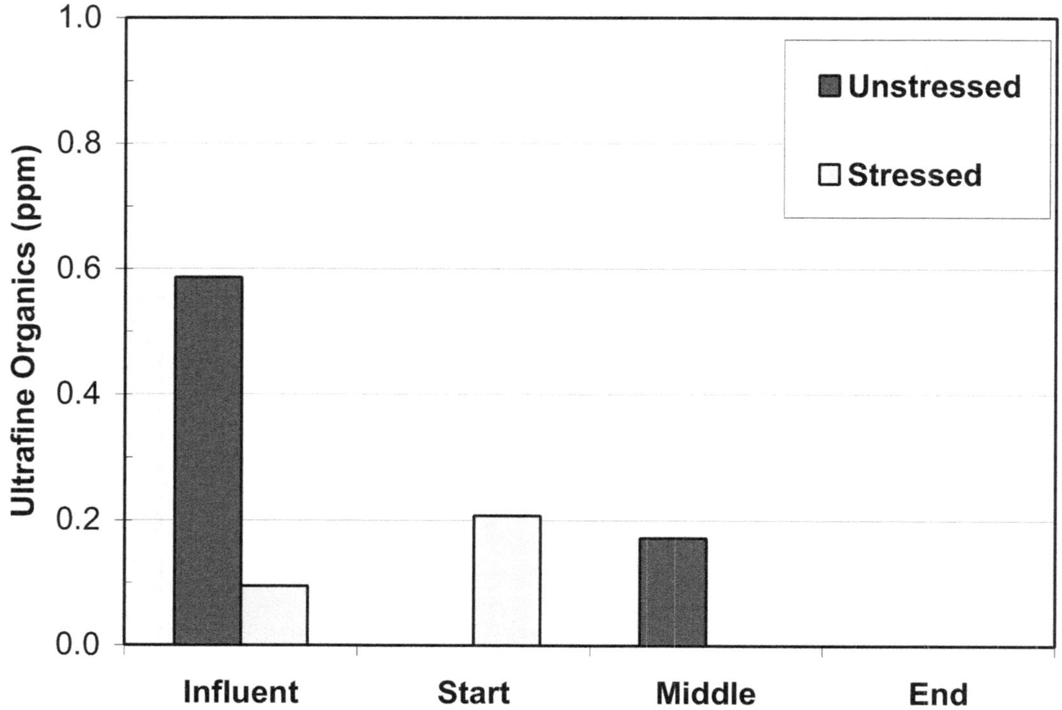

Figure 5.40 Concentration of TOC in the ultrafine fraction in the filter influent and effluent at the Southern Nevada water treatment plant during stressed and unstressed operating conditions

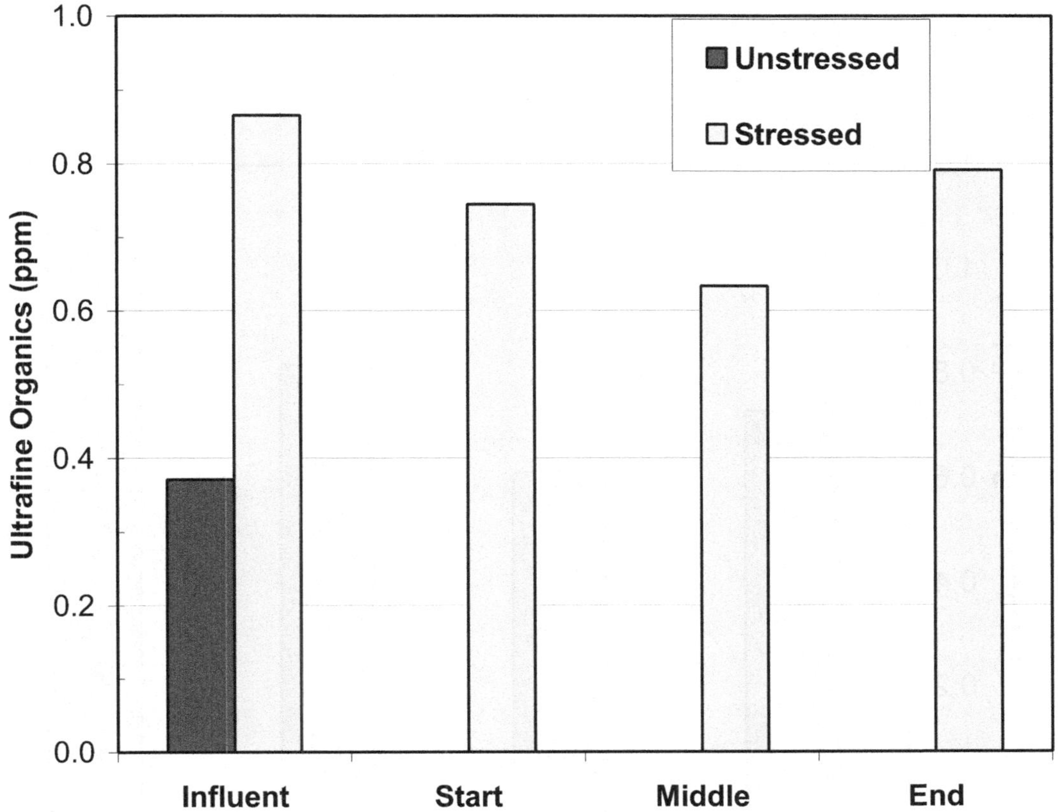

Figure 5.41 Concentration of TOC in the ultrafine fraction in the filter influent and effluent at the Johnson County water treatment plant during stressed and unstressed operating conditions

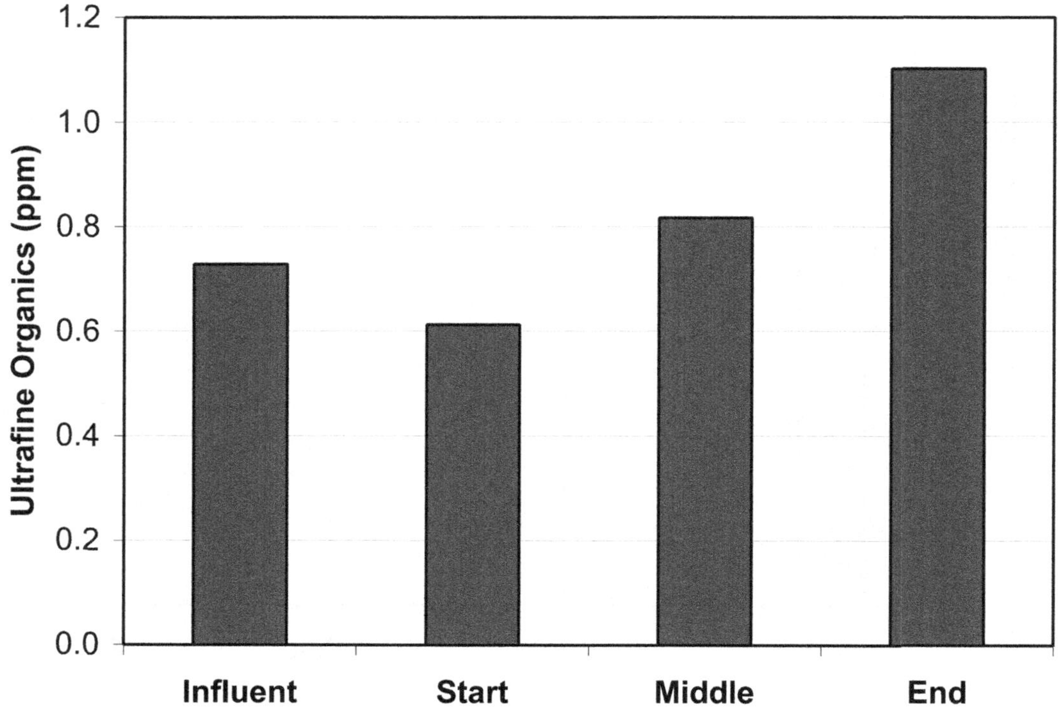

Figure 5.42 Concentration of TOC in the ultrafine fraction in the filter influent and effluent at the Buffalo Pound water treatment plant during stressed operating conditions

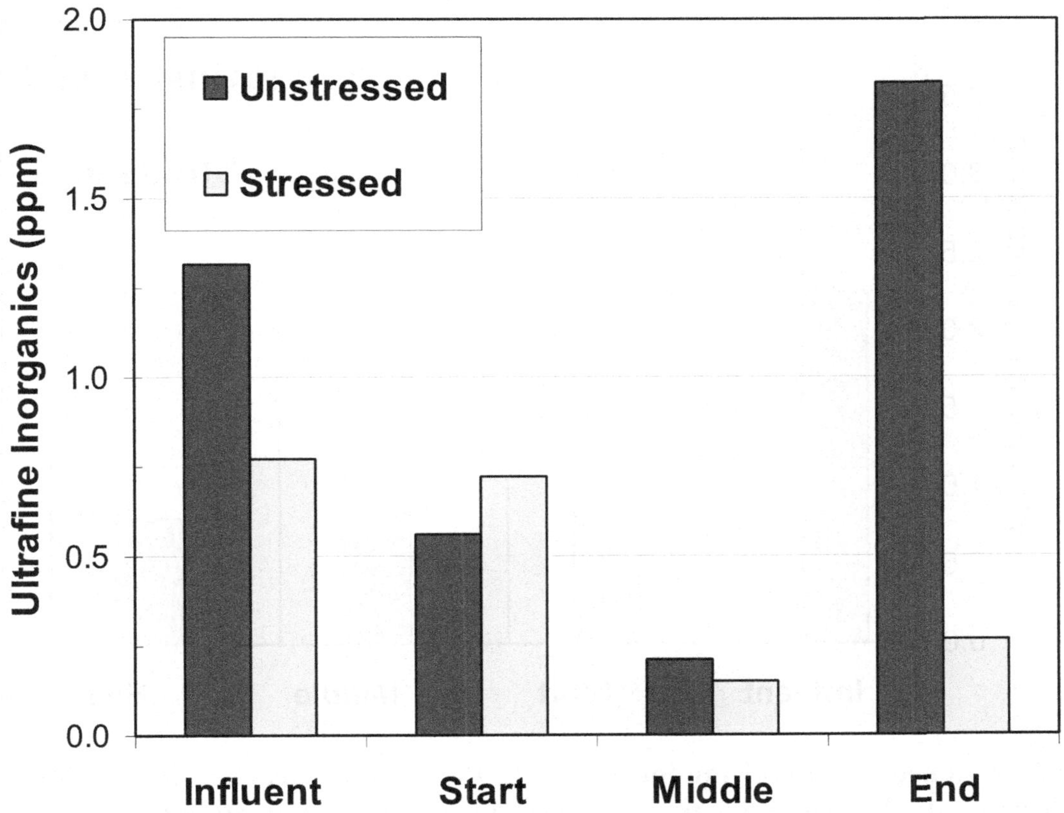

Figure 5.43 Concentration of the inorganics in the ultrafine fraction in the filter influent and effluent at the Milwaukee water treatment plant during stressed and unstressed operating conditions

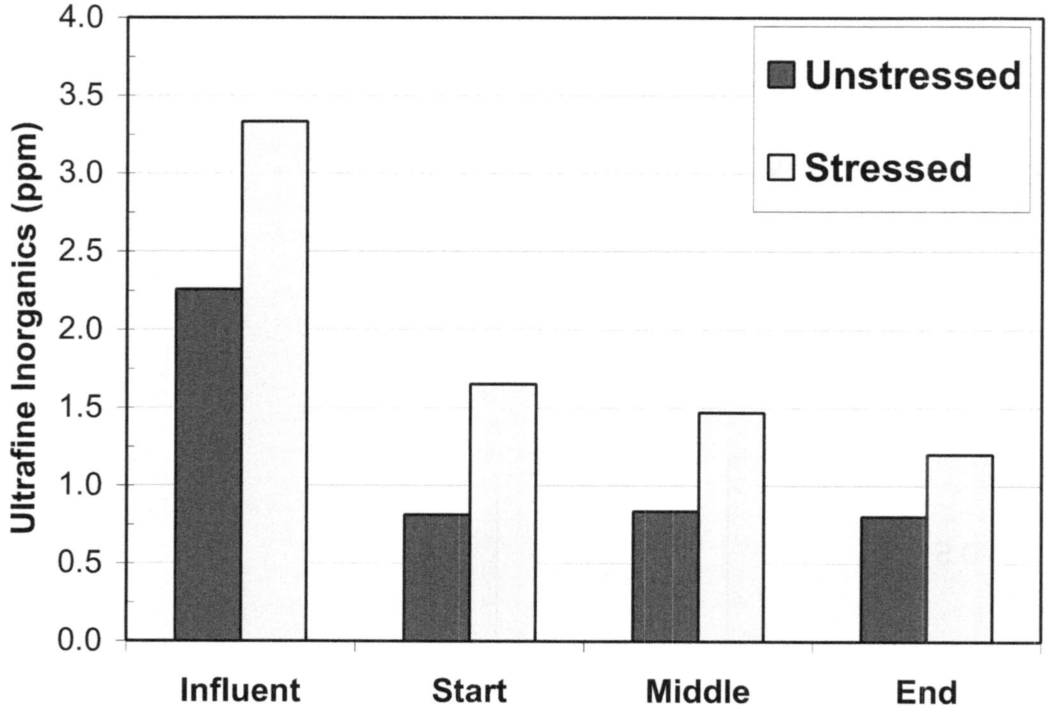

Figure 5.44 Concentration of the inorganics in the ultrafine fraction in the filter influent and effluent at the Modesto water treatment plant during stressed and unstressed operating conditions

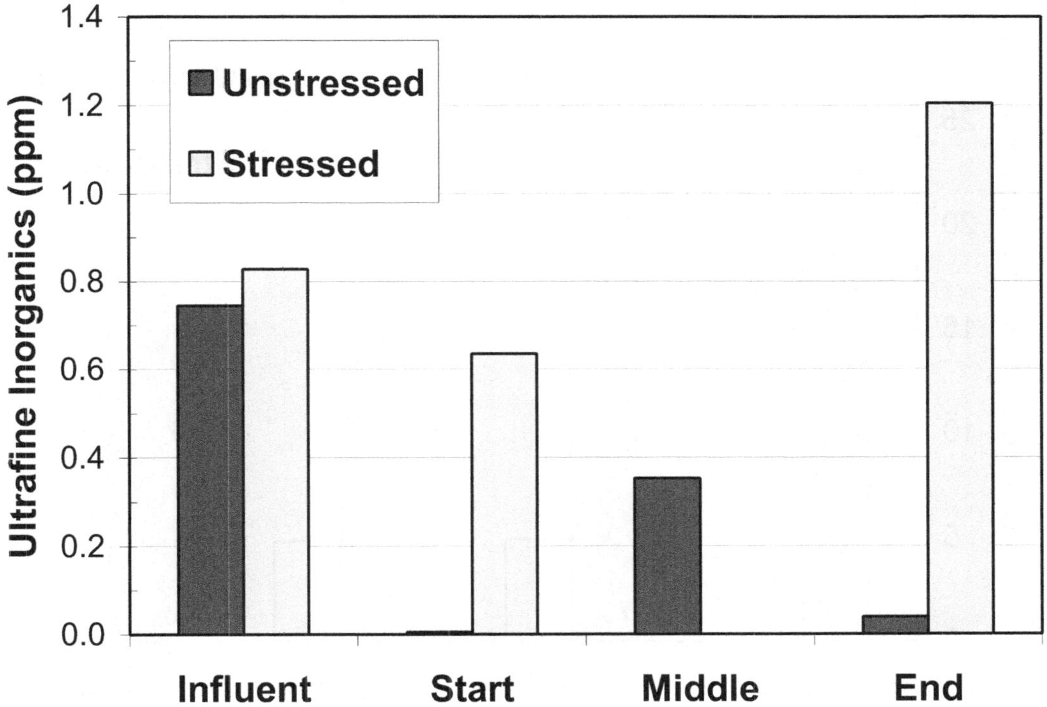

Figure 5.45 Concentration of the inorganics in the ultrafine fraction in the filter influent and effluent at the Northern Kentucky water treatment plant during stressed and unstressed operating conditions

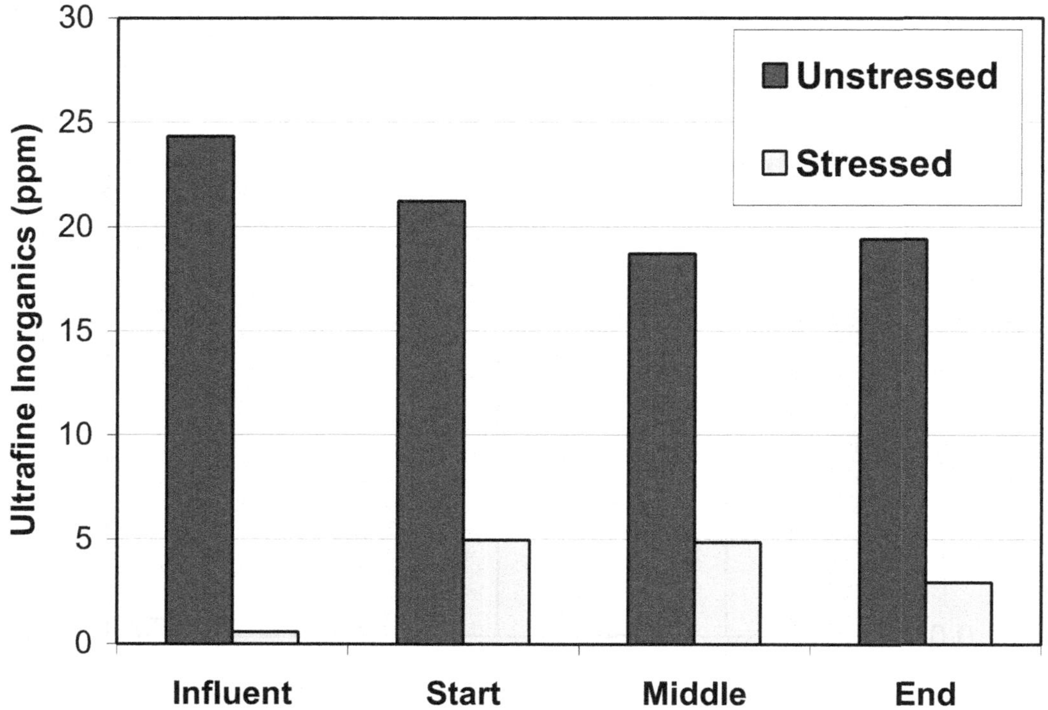

Figure 5.46 Concentration of the inorganics in the ultrafine fraction in the filter influent and effluent at the Houston water treatment plant during stressed and unstressed operating conditions

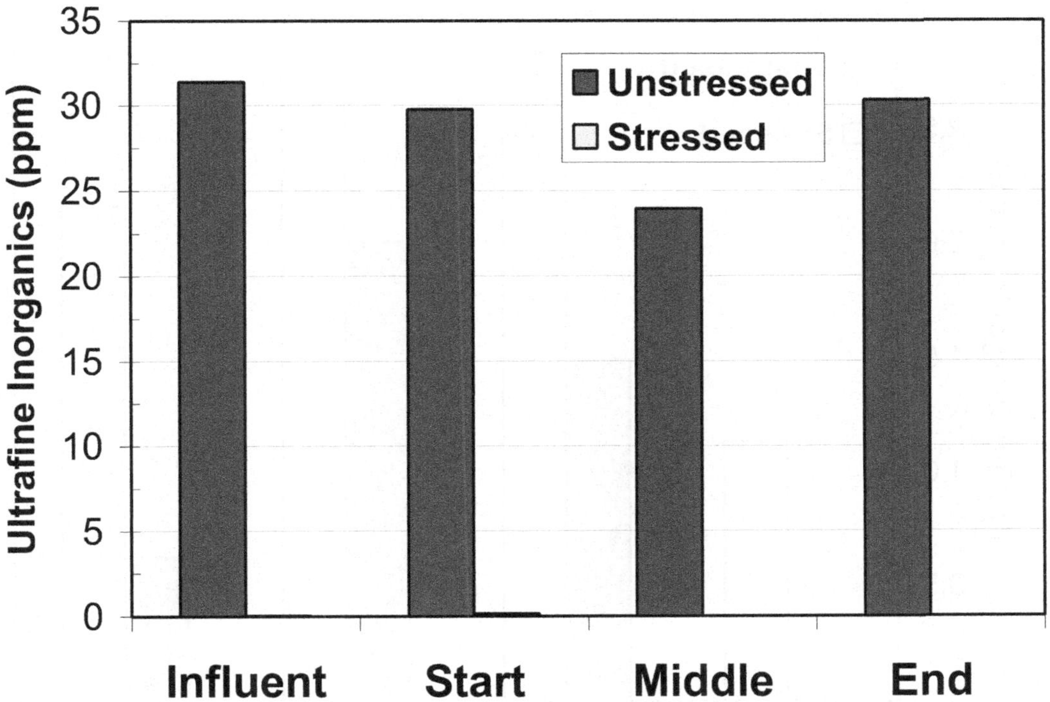

Figure 5.47 Concentration of the inorganics in the ultrafine fraction in the filter influent and effluent at the Cincinnati water treatment plant during stressed and unstressed operating conditions

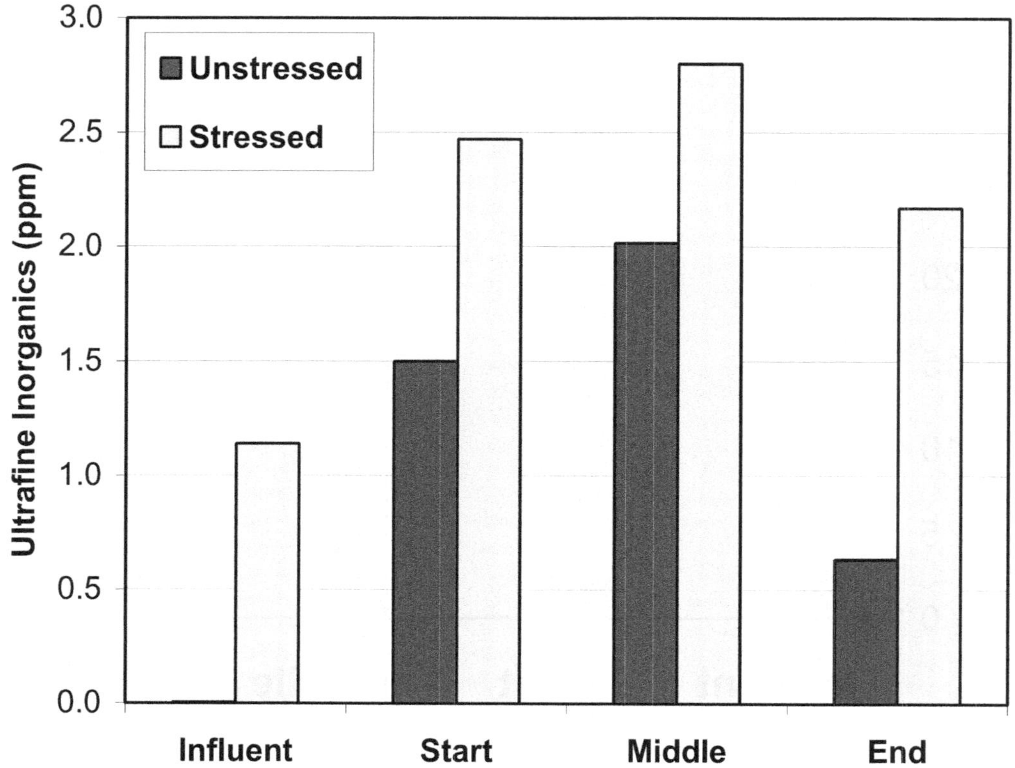

Figure 5.48 Concentration of the inorganics in the ultrafine fraction in the filter influent and effluent at the Southern Nevada water treatment plant during stressed and unstressed operating conditions

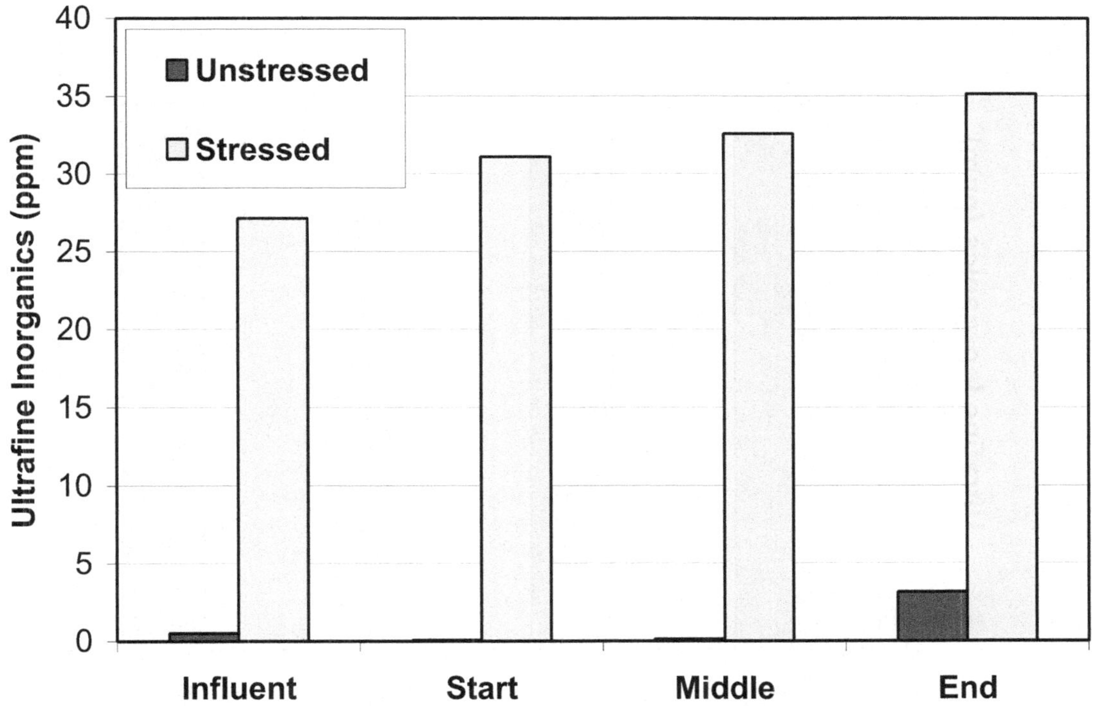

Figure 5.49 Concentration of the inorganics in the ultrafine fraction in the filter influent and effluent at the Johnson County water treatment plant during stressed and unstressed operating conditions

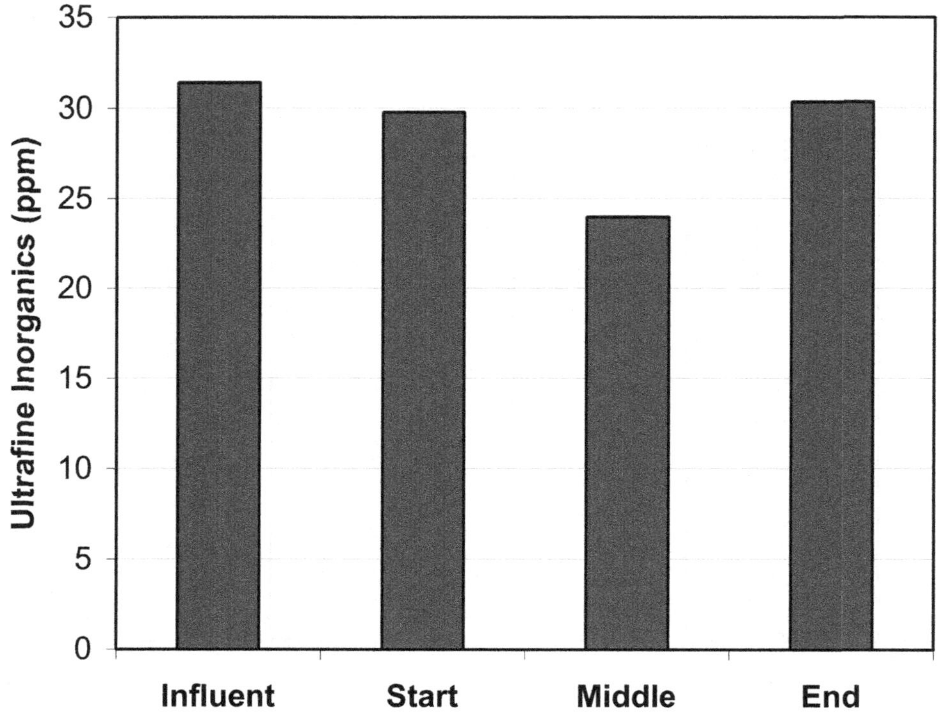

Figure 5.50 Concentration of the inorganics in the ultrafine fraction in the filter influent and effluent at the Buffalo Pound water treatment plant during stressed operating conditions

CHAPTER 6
CONCLUSIONS AND RECOMMENDATIONS

The overall objectives of this research were to characterize the inorganic and organic particle composition in filter influent and effluent streams representing a broad range of filtration and pretreatment practices, and contrast particle composition as a function of conditions of pretreatment, filter design and operation, and the raw water characteristics. In this chapter we summarize the most significant findings of this study and present our conclusions based on upon the reported data. Relevance of these results to the utilities is also discussed with respect to some of the pertinent conclusions. Additionally, we note our recommendations for future research based on the findings of this project.

KEY CONCLUSIONS AND RELEVANCE TO UTILITIES

- Traditional turbidity measurements do not reliably indicate the amount of organic and inorganic materials in the submicron size groups (i.e., ultrafine and dissolved +). Overall, the effluent turbidity did not correlate well with the amount of materials across all size fractions in the filter effluent when considering all of the utilities.
    - The inadequacy of the existing turbidimeters in detecting variations of organic and inorganic material in submicron size groups limits the degree of optimization that can be practiced in daily operation of facilities. While many utilities have developed site specific operational procedures to provide safe potable water for their customers, use of more sensitive instruments should be considered for monitoring performance of their filters. Currently, there are no online instruments available that allow continuous monitoring of particles down to the nanometer size range that are evaluated in this study. However, some of the available instruments such as laser turbidimeters can extend the detection limit to submicron particles and could be considered.
- Organic and inorganic materials in the filter influent and effluent are predominantly present in the dissolved + size range. Approximately 80% of the inorganic material and 70% of organic material in the influent are present in dissolved + size range. In the filter effluent, approximately 76% of the inorganic and organic material is present in the dissolved + size range. The dissolved + size fraction is largely composed of hardness cations and silica, in addition to residual amounts of metal coagulant species, specifically aluminum.
    - The high fraction of organic material present in the dissolved + size fraction suggests low concentrations, if any, of pathogens in the filter influent as well as effluent for the utilities investigated in this study. While the typical size of viruses is within the dissolved + and ultrafine size classes defined in this study, the disinfection process typically achieves a high degree of inactivation. The presence of inorganic constituents such as hardness cations, silica and residual amounts of metal coagulants are of some concern for some of the potential downstream processes, such as reverse osmosis. Precipitates of metal silicates could potentially foul the reverse osmosis. While the low concentrations of silica, aluminum and iron observed in this study are typically not a concern, careful

consideration should be given for downstream processes. The relatively high concentration of aluminum observed in the effluent at some utilities suggests that further optimization of pretreatment processes is possible.

- Materials in the ultrafine size fraction comprised approximately 14% to 20% of the total organic and inorganic material mass in the filter influent and were primarily composed of calcium, magnesium, silica, and organic materials. In the filter effluent, the ultrafine size fraction of these materials ranged from approximately 10 to 20%. A moderately strong correlation ($R^2 \sim 0.8$) existed between the amount of inorganic and organic materials in the ultrafine size fraction.
    - Material in the ultrafine size fraction, particularly for the influent is a concern with respect to filter headloss development. The generation of ultrafine particles within filter media observed at some of the utilities investigated needs further investigation. Ultrafine particle concentrations may be controlled by optimizing pretreatment upstream of filters, and by evaluating filter operating conditions to minimize shearing off of previously formed deposits. While the observed results indicate that the ultrafine mass fraction is typically low in the filter effluent, it is dependent on the stage of the filter run also, exhibiting higher concentrations at the beginning and end of the filter runs.
- There is significant variation among various utilities suggesting water quality, and type of treatment can significantly influence the size distribution of organic and inorganic material.
- The granular media filters did not significantly reduce the overall concentration of materials in the dissolved + size fraction. However, the observed degree of removal varied considerably, indicating that the ability of filters to remove particles in the nanometer size range will vary based on nanoparticle composition and influent chemistry.
- The types of materials entering and exiting granular media filters are diverse and dependent on the unique conditions of each plant.
- The effluent from the filters is mostly composed of materials in the dissolved + size fraction and are thus difficult to detect with conventional techniques.
- In several instances, higher concentrations of organic and inorganic material in the filter effluents compared to that of influent were observed. This suggests that smaller particles may aggregate, or re-suspend, following deposition in the filter media. Larger materials are more effectively removed at low filtration rates, while removal was less efficient or resulted in the generation of larger particle in the filter effluent at higher filtration rates.
- Among the pre-oxidation scenarios evaluated, pre-chlorination was most associated with the presence of larger size particles in the filter effluents. The longest filter run-times were observed at the pre-ozonated plants.
    - Since chlorination also results in formation of disinfection byproducts, utilities should consider delaying chlorination until after filtration, particularly, if it is evident that chlorination is affecting their filter effluent with respect to particle concentrations.
- Residual coagulant metal concentrations are effectively reduced during filtration although their presence in the filter effluent may contribute to the formation of cakes on membrane filters that may be placed downstream. No correlation was found between the loading rates of residual coagulants and filter run-time.

- The results indicated that in general, the coagulation conditions were well optimized at the facilities that were sampled in this study. However, in some instances, the relatively higher concentration of coagulant residuals suggests that further optimization is possible. Based on the mass removal of various organic and inorganic constituents, it appears that the residual coagulants may potentially contribute to a significant part of the headloss developed in the filter media.

- Filter run-times tend to increase with increasing Ca/TOC ratios, especially after a Ca/TOC value of 20. Additionally, filter run-times tended to decrease with increasing organic loading rates.
  - Further investigation of these results is required to understand the effects of organic material on filter performance. Based on these results, it appears that removal of organic material can improve the performance of filters.
- Complexation between organic and inorganic materials is likely and may ultimately result in the generation of larger particles in filter effluents.
- Longer filter run-times are generally achieved when the concentration of materials in the ultrafine size fraction are low. Particles in this size range may thus have the most significant impact on filter performance.

SUMMARY OF THE RESULTS

Overall Influent and Effluent Composition

- Inorganic and organic materials in the filter influent were predominantly present in the dissolved + and ultrafine size fractions.
- At a given time the filters removed little of the total influent organic (~10%) and inorganic (~3%) materials that were analyzed for during both the stressed and unstressed periods. Thus, materials that are retained in the filter build up slowly over time and do not result in a sudden increase in headloss. No one contributor could be identified as being particularly detrimental to the filter performance, although several constituents were found to be retained to a greater degree in the filter than others: coagulant residuals and manganese.
- Calcium, magnesium, and silica, comprised around 98% of the total mass (~ 50 ppm) of inorganic materials that were analyzed for in the filter influent and effluent at each of the different utilities. Following the hardness ions and silica, both iron and aluminum were the most prevalent elements in the filter influent and effluent.
- Residual coagulant species in the larger size fractions were removed by the filter and were mostly present in the dissolved + size fraction in the filter effluent.
- Concentrations of lead, copper, arsenic, and manganese in the filter influent and effluent were significantly lower than those for the other elements analyzed and were mostly present in the dissolved + size fraction.
- Several different types of large filter-clogging diatoms (up to 100 μm in size) were identified in the influent and effluent at some of the water treatment plants. It is likely that most of these diatoms originated from sediment accumulated in the raw water intake pipe and filter drain. This suggests that the product water quality may be impaired by contamination in the post-filtration system.

Operational Factors and Filter Performances

Pretreatment

- Pre-oxidation appears to shift materials to the larger particulate size fractions. For the plants that use pre-ozonation, approximately 50% of the total organic mass was in the three largest size fractions (i.e., settleable, filtrable or ultrafine) during the unstressed period. However, under stressed conditions the bulk of the TOC mass was in the dissolved + size fraction. Similarly, inorganics in the filter influent existed to a greater degree in the larger size fractions during unstressed conditions. These observations are likely a consequence of the definition of stressed and unstressed conditions. During stressed conditions, filter run-times tend to be shorter and coagulant doses higher. Higher coagulant doses would be required during periods of higher concentrations of natural organic matter in the raw water, leading to higher solids loadings on the filters.
- Following plants with pre-ozonation, facilities that practice pre-chlorination were observed to have high concentrations of particulate organic and inorganic materials with approximately 55% for organics and 40% for inorganics during the unstressed period.
- The most substantial fraction of inorganic and organic materials in the larger size fractions, mostly as ultrafines during the stressed period, was obtained at the plant where chlorine dioxide and potassium permanganate were used together as pre-oxidants.
- Manganese was generally found in the dissolved + fraction.

Coagulant Residuals

- No strong correlation between either of the residual coagulant species (Fe or Al) and filter run-times was evident, suggesting that at the observed concentrations of these species, other factors control the filter run time.
- The effluent concentrations of aluminum and iron generally fell below the maximum target values for these materials, 0.2 and 0.3 ppm for aluminum and iron, respectively.

Filter Operational Factors

- No correlation was found between filtration rate and the removal of organic materials. It appears that the shift of organic particles to the larger size classes may mask the effect of filtration rate on the removal of these particles. The removal of inorganic materials in larger size classes appears to be affected to certain extent by filtration rate. In general, facilities operating at higher loading rate had higher concentrations of inorganic material in the larger size classes, compared to that of influent..
- No significant difference between the removal efficiencies amongst the different media types was observed.

Feed Water Quality

- No significant differences were observed for the filter influent water quality and filter performance during the stressed and unstressed periods. This suggests that pretreatment and filter operation are optimized sufficiently to maintain optimal filter performance.

- Filter run-time did not directly correlate with influent organic carbon concentrations, attesting to the complexity of the problem and indicating that other factors such as influent solution chemistry and filter operating parameters (e.g., filter media used and filtration rate) must also be considered.
- Filter run times increased with increasing Ca/TOC ratios in a logarithmic fashion once the Ca/TOC ratio exceeded a value of 20.

Rapid and Slow Filtration

- The major difference between rapid and slow sand filter effluent quality was the fraction of material in the dissolved + size class. The organic matter was predominantly present in the dissolved + size fraction in the influent and effluent of rapid sand filters, with approximately 67% and 73% respectively. For the slow sand filter, the percentage of organic matter was approximately 38% and 32% in the influent and effluent, respectively.
- The dissolved + size fraction of inorganic material was about 80% and 81% in the influent and effluent of rapid sand filters. However, the dissolved + size fraction was higher in slow sand filter, at approximately 87% and 96% in the influent and effluent, respectively.

Relationship Between Turbidity and Size Fractions

- There was no relationship between the relative size distributions of organic and inorganic materials in the filter influent and effluent and the measured turbidity during both the stressed and unstressed periods.

Ultrafine Fraction

- There was an almost linear relationship between the concentration of organic and inorganic materials in the ultrafine size fraction. The ratio of inorganic and organic material concentrations in the ultrafine size range was approximately 1.2.
- No direct correlation was observed between the concentrations of materials in the ultrafine size fraction and filter run time. However, at the data indicated that the potential for shorter filter runs at higher loading rates of organic (> 200 mg/min/ft^2) and inorganic material (> 20 mg/min/ft^2) is higher.
- In most cases, the TOC concentration in the ultrafine size fraction increased in the filter effluent. The reason for this is not clear although several possibilities can be hypothesized. For instance, the deposits of smaller organic particles may be eluting the filter in the ultrafine size fraction. Adsorption of organics to ultrafine colloids is also a possible scenario. Potential complexation reactions between dissolved organic and inorganic materials that shift macromolecular organic matter into the particulate range are also a possibility.
- In contrast to the ultrafine organic material, the concentration of inorganic material in the ultrafine size fraction decreased in the effluent at most of the plants for both stressed and unstressed periods. This may be due to the more favorable adhesion of these materials to the filter media compared to that for the organics.

FUTURE RESEARCH RECOMMENDATIONS

This project examined the characteristics, size and origin, of materials in filter influent and effluent streams at a variety of full-scale water treatment plants. Based on the results of this study, further research on the following issues is suggested.

- The focus of this project was the characterization of particles in filter effluents. To that extent, the filter influent and effluent water quality at various facilities was investigated. To obtain a better understanding of the filter effluent water quality as a function of raw water quality and treatment processes, the water quality through various treatment processes up to the filtration process should also be investigated. For instance, the amount of organic and inorganic material in various size classes is likely to vary after each unit treatment process. If a particular size class (such as the ultrafine fraction) is affecting the filtration process, the amount of material in that size class may be limited by optimizing processes upstream.
- Results from this study indicated the potential for significant changes to organic and inorganic material size distribution within the filter media. For instance, in several utilities, the concentrations of organic material in the ultrafine size fraction appeared to be higher in the effluent, compared to the influent. Further investigation of the balance of organic matter throughout multiple filtration cycles is required to explain these observations.
- The effect of calcium and TOC concentrations ratio on the filter run times needs further investigation.
- The lack of any observed correlation between turbidity and concentration of organic and inorganic materials in the submicron size classes should be investigated further. While this may signal limitations in the use of turbidity to monitor these smaller size classes, more advanced instruments such as laser turbidimeters should be evaluated.
- The smallest size class evaluated in the current study was limited to dissolved + size class, defined as material smaller than 30 nm. Given that the bulk of organic and inorganic material is present in this size class, various means of further declassification of this size class should be investigated.
- This study primarily focused on organic and inorganic material in the filter influent and effluents. However, given that substantial material was present in the dissolved + size range, and that several particles of biological origin may be present within such range (for example, viruses, fragments of cells such as cell walls, proteins, etc.), future studies should extend the current study to identify various components of organic material in the filter influent and effluent.
- Greater resolution of materials smaller than 30 nm is particularly important given the need to track removal of viruses in filters. A comparison of removal of viruses of different varieties with inorganic and other organic particles is needed.
- This study was conducted at several full scale facilities. For public safety, the sampling schedule with respect to filter run stages was based on the utilities' practices. Additional information on filter breakthrough (such as the sequence of particle size classes that break through the filter) should be gathered.
- Variability between facilities and through time at full scale, tend to obscure general trends in filter behavior. Investigations at pilot or lab-scale under highly controlled

conditions will be required to arrive at true mechanistic understanding of the interactions between particle size classes within filters, the role of coagulant type on the presence of particles in filter effluents, and the bounds for filter optimization in removing very small particles.

- A strong correlation between organic and inorganic material in the ultrafine size fractions was observed in this study. While it is postulated that interactions between inorganic particles and organic material might have led to this correlation, further investigation is required to identify the nature of interactions.
- The effect of organic and inorganic materials in various size classes on processes downstream should be investigated. For instance, some of the utilities investigated had granular activated carbon. Extension of the analyses conducted in this study on the effluent of processes downstream of filters would be beneficial to many utilities.

APPENDIX A

Field Sampling Protocol

STEP 1 – SAMPLE PRESERVATION

Sample Preservation

Most of the samples will require the addition of a preservative. The preservative will be a small volume of liquid in the closed bottle. The addition of preservative is typically performed at Rice University, but in some instances it may be necessary to add the preservative in the field. If the preservative is in the bottle, do not clean the bottle or pour the preservative out. If the preservative is not in the bottle, you will need to perform the following steps to each of the 5 types of bottles.
Glass containers are for total organic carbon (TOC) analysis. Sodium azide will be added to these containers at a concentration of 1 mg/L of sample. This prevents microbiological activity.

500 mL clear glass bottles

- Prepare a dilution of the commercial sodium azide 5 percent w/v product. Pipet 2 mL of the 5 percent w/v sodium azide solution to a 100 mL volumetric flask and dilute to 100 mL.
- Pipet 0.50 mL of the diluted sodium azide solution and add to the 500 mL sample containers.

250 mL amber glass bottles

- Pipet 0.25 mL of the diluted sodium azide solution and add to the 250 mL sample containers.

Plastic containers are for metals analysis. Nitric acid will be added to these containers at a concentration of 1.5 mg/L of sample. The pH of the sample should be reduced to less than pH 2.

250 mL amber glass bottles

- Pipet 0.4 mL of the dilute nitric acid solution and add to the 250 mL sample containers.

In addition to sample preservation, 0.45 um membrane filters will need to be conditioned for metals analysis. Each filter must be washed with 50 mL of distilled water. The water should be filtered to remove any metals and solution wasted. The protocol for filter preparation can be found in Standard Methods 3030 B. Filtration for Dissolved and Suspended Metals. Each filter must be weighted before water for filtration.

STEP 2 - FIELD SAMPLING

Filter Influent Sample

Number of Locations:
3 (3 separate filters for full sampling) / 1 (1 filter for reduced sampling)

Number of samples at each location:
Qty. 1 – 4 L sample
Qty. 1 – 500 mL sample

Bottles:
4 L clear glass bottle
500 mL clear glass bottle, with preservative

Sample type: Composite

Procedure: Collect approximately 500 ml of filter influent every 4 hours over the course of filter run and pour it into the 4 L sample container. Collect approximately 125 mL of filter influent every 4 hours over the course of filter run and pour it into the 500 mL sample container. The composite sample volume and frequency should be adjusted on site for variations in filter run length.

Filter Effluent Sample

Number of Locations:
3 (3 separate filters for full sampling) / 1 (1 filter for reduced sampling)

Number of samples at each location:
Qty. 3 – 4 L sample (1. filter ripening, 2. steady state and , 3. just before backwash)
Qty. 3 – 1.5 L sample (1. filter ripening, 2. steady state and, 3. just before backwash)

Bottles: 4 L clear glass bottle
500 mL clear glass bottle, with preservative

Sample type: Grab

Procedure: After two bed volumes of water has passed through the filter, fill the 4 L and 500 mL containers with the sample, headspace free. Repeat midway through the run once turbidity values remain low for an extended period of time. Filter effluent sampling should be performed a third time prior to backwash.

Backwash Water Sample

Number of Locations: 1

Qty. 1 – 4 L samples

Bottles: 4 L clear glass bottle

Sample type: Composite

Procedure: Collect approximately 1 L of backwash water following the displacement of half a bed volume of backwash water. This should be slightly after the dirtiest water begins to enter the backwash troughs. After that collection, collect approximately 400 mL every 2 minutes until the backwash is complete. (composite sample volume and frequency should be adjusted on site for variations in filter backwash period.) Perform this for just one filter.

STEP 3 – SAMPLE PROCESSING

Filter Influent Fractionation

<u>500 mL Sample</u>

- No fractionation step is required for this sample.
- Label this sample as follows and in accordance with Step 4. "In – F_ - G". The label should show that the preservative "Azide" is present.

<u>4 L Samples</u>

Sedimentation Sample

- Swirl or stir sample to suspend particles.
- Remove 1 L of the sample and pour approximately 1 L glass beaker.
- Set the sample aside and allow particles to settled for 1 minute.
- Remove the top 500 mL with syringe, being careful not to suspend particles that may have settled.
- Swirl the remaining 500 mL to suspend particles and pour 250 mL into a 250 mL HPDE plastic bottle and 250 mL into a 250 mL amber, glass bottle.
- Label this sample as follows and in accordance with Step 4. "In – F__ - S - __." The label should show that the preservative "Azide" is present.

Filtered Sample

- Swirl the 4 L sample to suspend particles.
- Pour 500 mL into the volumetric cylinder and filter the 500 mL volume through the 0.45 um washed membrane filter. Perform this with 2 separate filters, if necessary.
- Pour 250 mL of the filtered solution into the 250 mL amber glass bottle and label the bottle in accordance with Step 4. "In – F__ - F - __." The label should show that the preservative "Azide" is present.
- Pour 250 mL of the filtered solution into the 250 mL plastic bottle and label the bottle in accordance with Step 4. "In – F__ - F - __." The label should show that the preservative "acid" is present.

Filters

- All filters should be collected and placed in a single petri dish.
- The dish should be taped shut and label the container in accordance with Step 4. "In – F__ - F - __."

Filter Effluent Fractionation

Filter effluent fractionation is the same as that performed for filter influent. The only modification is labeling. Only labeling instructions are provided in this discussion.

<u>500 mL Sample</u>

- Follow protocol for Filter Influent Fractionation of 500 mL Sample.

- Label this sample as follows and in accordance with Step 4. “Ef – F_ - G.” The label should show that the preservative “Azide” is present.

4 L Sample

Sedimentation Sample

- Follow protocol for Filter Influent Fractionation of 4 L Sedimentation Sample.
- Label this sample as follows and in accordance with Step 4. “Ef – F__ - S - __.” The label should show that the preservative “Azide” is present.

Filtered Sample

- Follow protocol for Filter Influent Fractionation of 4 L Filtered Sample.
- Label the bottle in accordance with Step 4. “Ef – F__ - F - __.” The label should show that the preservative “acid” is present.

Filters

- All filters should be collected and placed in a single petri dish.
- The dish should be taped shut and label the container in accordance with Step 4. “Ef – F__ - F - __.”

Filter Backwash Water Fractionation

- Collect a 4L water sample
- Swirl the 4 L sample to suspend particles.
- Pour 250 mL of the solution into the 250 mL amber glass bottle and label the bottle in accordance with Step 4. “Backwash”. The label should show that the preservative “Azide” is present.
- Pour 250 m L of the filtered solution into the 250 mL plastic bottle and label the bottle in accordance with Step 4. “Backwash”. The label should show that the preservative “acid” is present.

STEP 4 – SAMPLE LABELS AND CHAIN OF CUSTODY FORM

Sample identification should follow the consistent pattern provided in this section. In addition to proper sample identification, a laboratory notebook should be kept that includes the preparation details of all samples.

The labeling system consists of four values separated by a hyphen. Each of the four values are discussed further below and in Figure 1. The four sections represent the following parameters:

1. The process step at which the sample was taken
    - Filter Influent (IN)
    - Filter Effluent (EF)
    - Backwash (Backwash)
2. The filter from which the sample was taken
    - Filter Number (F#)
3. The sample preparation step

- Grab or Composite Sample (Grab)
- Settled Sample (S)
- Filtered Sample (S)

4. The portion of the filter run the sample was collected
   - Beginning of the filter run (B)
   - Middle of the filter run (M)
   - End of the filter run (E)

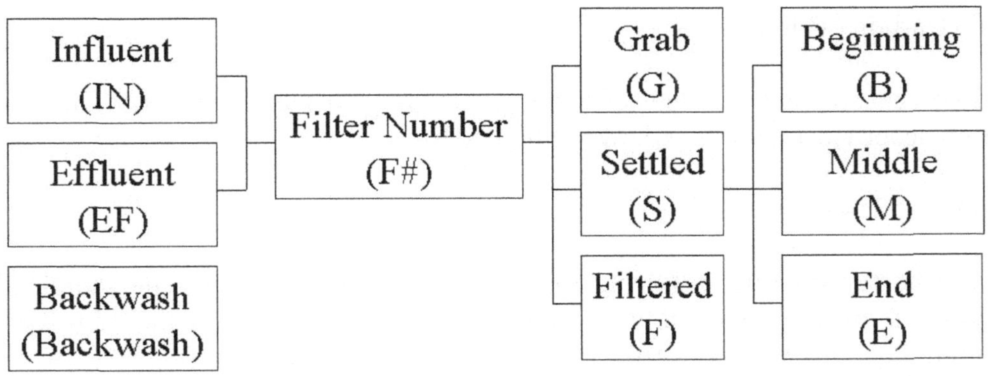

Figure A1. Labeling protocol and naming scheme for water samples collected on-site.

Based on this labeling protocol, a filter effluent sample from Filter No. 13 that is filtered and taken at the end of the filter run would be labeled "EF – F13 – F – E." Note that there will be two samples labeled the same. One sample will be in a glass container and the other in a plastic container. They will undergo different analyses, the glass bottle TOC, and the plastic bottle, metals.

The overall sampling procedure is illustrated in Figures A2, A3 and A4.

STEP 5 – ONSITE ANALYSIS

Some onsite analysis should be performed prior to shipping of the chemicals. Onsite analysis should be performed with the water remaining in the 4 L sample container. E.g. the number of influent samples that should be analyzed are three as that is the number of 4 L containers that will be filled during sampling. The analyses should be limited to only those that can be performed with equipment at the plant onsite and those that are already being sent to an outside laboratory for compliance purposes to limit the burden on the plant laboratory staff. The following analytical procedures should be performed:

Hydrogen Ion Concentration (pH)

Hydrogen ion concentration should be ran on filter influent, effluent, and backwash water. Hydrogen ion should be measured with a pH probe and in accordance with Standard Method 4500-H^+.

Temperature

Temperature should be run on filter influent, effluent, and backwash water. Temperature should be measured in accordance with Standard Method 2550.

Turbidity

Turbidity should be run on filter influent, effluent, and backwash water. Turbidity should be measured in accordance with Standard Method 2130.

Total Organic Carbon (TOC)

TOC should be run on filter influent, effluent, and backwash water. TOC should be measured and sampled in accordance with Standard Method 5310.

UV Absorbance at 254 nm (UVA_{254})

UVA_{254} should be run on filter influent, effluent, and backwash water. The sample should be filtered with a 0.45 um filter and analyzed in a 1 cm quartz cell. UVA_{254} should be measured in accordance with Standard Method 5910.

True Color

True color should be run on filter influent, effluent, and backwash water. The sample should be filtered with a 0.45 um filter and analyzed. True color should be measured in accordance with Standard Method 2120.

Hardness

Hardness should be run on filter influent, effluent, and backwash water. Hardness should be measured in accordance with Standard Method 2340.

Alkalinity

Alkalinity should be run on filter influent, effluent, and backwash water. Alkalinity should be measured in accordance with Standard Method 2320.

Conductivity

Conductivity should be run on filter influent, effluent, and backwash water. Conductivity should be measured in accordance with Standard Method 2510.

Particle Counts

Particle counts greater than 2 um diameter should be run on filter influent, effluent, and backwash water. Particle counts should be measured in accordance with Standard Method 2560.

Total Chlorine

Total chlorine should be run on filter influent, effluent, and backwash water. Total chlorine should be measured in accordance with Standard Method 4500-Cl.

Free Chlorine

Total chlorine should be run on filter influent, effluent, and backwash water. Total chlorine should be measured in accordance with Standard Method 4500-Cl.

Biological Characterization

Heterotrophic Plate Count (HPC) and total coliforms should be determined on filter influent, effluent, and backwash water. HPC should be measured in accordance with Standard Method 9215. Total coliforms should be measured in accordance with Standard Method 9221, 9222, or 9223.

Disinfection Byproducts (DBPs)

If during the same period (1 week before or 1 week after testing) disinfection byproducts (trihalomethanes or haloacetic acids) are sampled in the distribution system, these values should be reported. The analysis of disinfection byproducts should be the responsibility of the utility.

STEP 6 – SHIPMENT

Samples should be wrapped in bubble wrap and ice packets and ice should be added to the coolers. Sample labels should be covered with clear backing tape to prevent smearing of the ink and prevent the removal of the sample label.

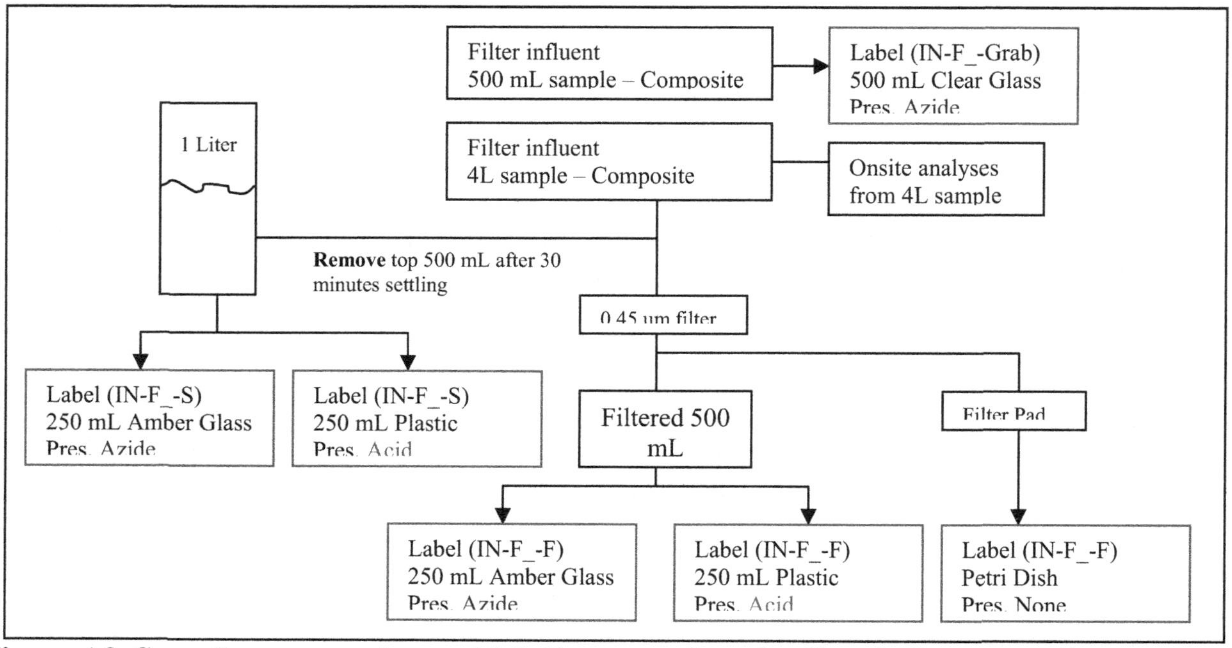

Figure A2. Sampling, processing and labeling procedure for filter influent samples.

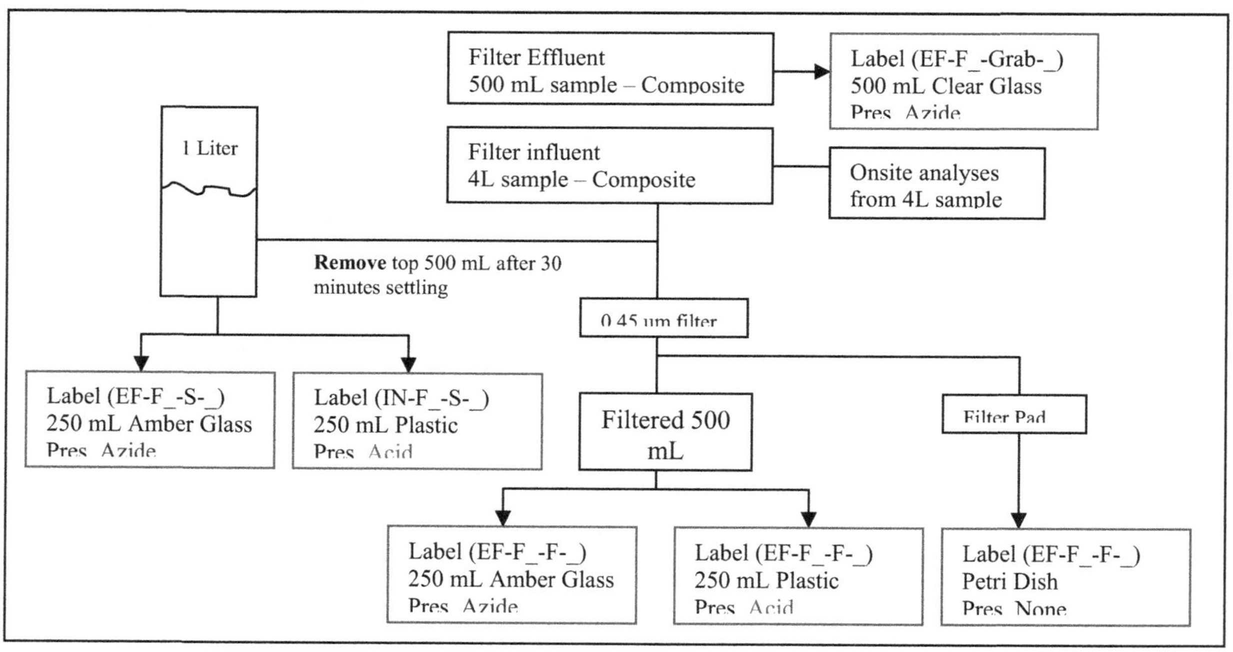

Figure A3. Sampling, processing and labeling procedure for filter effluent samples.

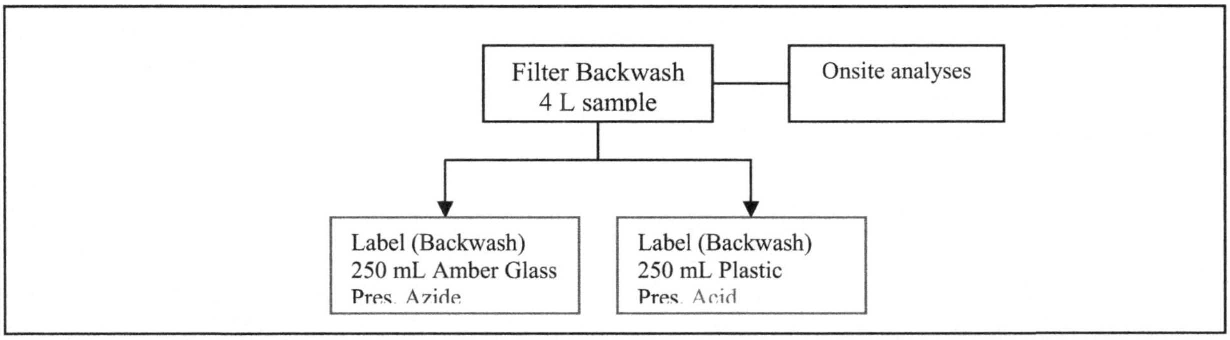

Figure A4. Sampling, processing and labeling procedure for backwash water samples.

REFERENCES

Al-Ani, M. Y. 1986. Giardia cysts from low turbidity waters by rapid rate filtration. *Journal American Water Works Association*, 78 (5): 66.

Amirtharajah, A. and Wetstein, D. P. 1980. Initial degradation of effluent quality during filtration. *Journal American Water Works Association*, 72 (9): 518.

Amirtharajah, A. 1993. Optimum backwashing of filters witth air scour: a review. *Water Science and Technology*, 27 (10): 195.

AwwaRF-IWSA. 1998. Treatment process selection for particle removal. International Water Supply Association American Water Works Association, Denver.

Becker, W. C. and O'Melia, C. R. 2001. Ozone: its effect on coagulation and filtration. *Water Science and Technology: Water Supply*, 1 (4): 81.

Becker, W. C., O'Melia, C. R., Au, K. and Young, J. S. 2004. Using oxidants to enhance filter performance. AWWARF, Denver.

Blackburn, B.G., Craun, G.F., Yoder, J.S., Hill, V., Calderon, R. L., Chen, N., Sherline, L. H., Levy, D.A., Beach, M. J., Surveillance for waterborne-disease outbreaks associated with drinking water – United States, 2001-2002. In: Surveillance Summaries, October 22, 2004. Morbidity and Mortality Weekly Report 2004; 53 (No. SS-8)

Boccelli, D. L., Small, M. J. and Diwekar, U. M. 2004. Treatment plant design for particulate removal. *Journal American Water Works Association*, 96 (11): 77.

CFR (Code of Federal Register). 2002. Title 40, Part 143

Chang, S. D. and Singer, P. C. 1988. The impact of ozonation on particle stability and the removal of TOC and THM precursors. *Journal American Water Works Association*, 71

Cleasby, J. L., Williamson, M. M. and Baumann, E. R. 1963. Effect of filter rate changes on water quality. *Journal American Water Works Association*, 55 (6): 869.

Cohen, R. D. and Probstein, R. F. 1986. Colloidal fouling of reverse osmosis membranes. *Journal of Colloid and Interface Science*, 114 (1): 194.

Conley, W. R. 1972. High rate filtration. *Journal American Water Works Association*, 64 (3): 203.

Crittenden, J.C., et al., 2005. Water Treatment Principles and Design. 2nd ed., Hoboken: John Wiley & Sons, Inc. New York.

Cutler, D. and Miller, G. 2005. The role of public health improvements in health advances: the twentieth-century United States. *Demography*, 2005. 42 (1): 1-22.

D'Antonio, R. G., Winn, R. E., Taylor, J. P., Gustafson, T. L., Current, W. L., Rhodes, M. M., Gary, G. W. and Zajac, R. A. 1985. A waterborne outbreak of Cryptosporidiosis in normal hosts. *Ann. Int. Medicine*, 103 886.

Dai, X. and Hozalski, R. M. 2002. Effect of NOM and biofilm on the removal of Cryptosporidium oocysts in rapid filters. *Water Research*, 36 3523.

David, C. and Grant, M. 2005. The Role of public health improvements in health advances: the twentieth-century united states. *Demography*, 42(1): 1-22

DiBernardo, L. and Cleasby, J. L. 1980. Declining rate versus constant rate filtration. *Journal of Environmental Engineering*, 106 (EE6): 1023.

Driscoll, C. T. and Letterman, R. D. 1988. Chemistry and fate of Al III in treated drinking water. *Journal of Environmental Engineering Division ASCE*, 114 (1): 21.

Dugan, N. R. and Williams, D. J. 2004. Removal of Cryptosporidium by in-line filtration: effects of coagulant type, filter loading rate and temperature. *Journal of Water Supply: Research and Technology*, 55 (1): 1.

Edwald, J., Tobiason, J. E., Udden, C. T., Kaminski, G. S., Dunn, H. J., Galant, P. B. and Kelley, M. B. 2003. Evaluation of the effect of recycle of waste filter backwash water on plant removals of Cryptosporidium. *Journal of Water Supply: Research and Technology*, 52 (4): 243.

Edzwald, J. K., Tobiason, J. E., Parento, L. M., Kelley, M. B., Kaminski, G. S., Dunn, H. J. and Galant, P. B. 2000. Giardia and Cryptosporidium removals by clarification and filtration under challenging conditions. *Journal American Water Works Association*, 92 (12): 70.

Emelko, M. B., Huck, P. M. and Slawson, R. M. 1999. Design and operational strategies for optimizing Cryptosporidiumm removal by filters. *AWWA Water Quality & Technology Conference*. Denver:

Emelko, M. B. 2003. Removal of viable and inactivated Cryptosporidium by dual and trimedia filtration. *Water Research*, 37 2998.

Farizoglu, B., Nuhouglu, A., Yildiz, E. and Keskinler, B. 2003. The performance of pumice as a filter bed material under rapid filtration conditions. *Filtration and Separation*, 40 (3): 41.

Fitzpatrick, C. S. B. 1998. Media properties and their effect on filter performance and backwashing. *Water Science and Technology*, 38 (6): 105.

Fitzpatrick, C. S. B., Campbell, A. T. and Cable, C. J. 1999. The effect of filtration rate changes on filtrate particle counts. *AWWA Water Quality & Technology*. Denver:

Fox, K. R. and Lytle, D. A. 1996. Milwaukee's crypto outbreak: investigation and recommendations. *Journal of the American Water Works Association*, 88 (9): 87.

Fox, K. R. 2001. *Microbial removal by filtration processes* in 'Controlling disinfection by products and microbial contaminants in drinking water' Eds. Clark, R. M., and Boutin, B., K.. USEPA, Cincinnati, Ohio.

Gaubert, J. Y., Piet, C., Gatel, D., Bonne, P., Hutchinson, J. and Cavard, J. 2002. The effect of preozonation on microorganisms and particle removal. *Water Science and Technology*, 41 (7): 9.

Gorczyca, B. and London, D. 2003. Characterization of particles in slow sand filtration at North Caribou Water Treatment Plant. *Water Quality Research Journal of Canada*, 38 (1): 153.

Grout, H., Wiesner, M. R. and Bottero, J.-Y. 1999. Analysis of colloidal phases in urban stormwater runoff. *Environmental Science & Technology*, 33 831.

Haas, C. N., French, K., Finch, G. R., and Guest, R. K. 2001. *Data Review on the Physical/Chemical Removal of Cryptosporidium*. AwwaRF, Denver, CO.

Hargesheimer, E. E., Lewis, C. M. and Yentsch, C. M. 1992. Evaluation of Particle Counting as a Measure of Treatment Plant Performance. AWWARF, Denver.

Hargesheimer, E. E. and Lewis, C. M. 1995. A Practical Guide to On-Line Particle Counting. AWWARF, Denver.

Hargesheimer, E. E., McTigue, N. E. and Lewis, C. M. 2000. Fundamentals of drinking water particle counting. AWWARF, Denver.

Hatukai, S., Ben-Tzur, Y. and Rebhum, M. 1997. Particle counts and size distribution in system design for removal of turbidity by granular deep bed filtration. *Water Science and Technology*, 36 (4): 225.

Hayes, E. B., Matte, T. D., O'Brien, T. R., McKinley, T. W., Logsdon, G. S., Rose, J. B., Ungar, B. L. P., Word, D. M., Pinsky, P. F., Cummings, M. L., Wilson, M. A., Long, E. G., Hurwitz, E. S. and Juranek, D. D. 1989. Large community outbreak of Cryptosporidiosis due to contamination of a filtered public water supply. *New England Journal of Medicine*, 320 1372.

Hendricks, D. W. 2000. Biological Particle Surrogates for Filtration Performance Evaluation. AwwaRF, Denver, Colorado.

Hilmoe, D. J. and Cleasby, J. L. 1986. Comparing constant rate and declining rate direct filtration of a surface water. *Journal American Water Works Association*, 78 (12): 26.

Hong, S., Elimelech, M. 1997. Chemical and physical aspects of natural organic matter (NOM) fouling of nanofiltration membranes. *Journal of Membrane Science*, 132 159.

Horn, J. B., Hendricks, D. W., Scanlan, J. M., Rozelle, L. T. and Trnka, W. C. 1988. Removing Giardia cysts and other particles from low turbidity waters using dual stage filtration. *Journal American Water Works Association*, 80 (2): 68.

Hsu, B., Huang, C.-P. and Pan, J. R. 2001. Filtration behaviors of Giardia and Cryptosporidium – ionic strength and pH effects. *Water Research*, 35 (16): 3777.

Hsu, B. and Yeh, H. 2003. Removal of Giardia and Cryptosporidium in drinking water treatment: a pilot scale study. *Water Research*, 37 1111.

Huck, P. M., Coffee, B. M., Emelko, M. B., Maurizio, D. D., Slawson, R. M., Anderson, W. B., Oever, J., Douglas, I. and O'Melia, C. R. 2002a. Effects of filter operation on Cryptosporidium removal. *Journal American Water Works Association*, 94 (6): 97.

Huck, P. M., Emelko, M. B., Coffey, B. M., Maurizio, D. D. and O'Melia, C. R. 2002b. Filter operation effects on pathogen removal. AwwaRF, Denver.

Hudson, H. E. 1959. Declining rate filtration. *Journal American Water Works Association*, 51 (11): 1455.

Ives, K. J. and Sholji, I. 1965. Research on Variables Affecting Filtration. *Journal of the Sanitary Engineering Division, American Society of Civil Engineers*, 91 (SA4): 1.

Joseph, C., Hamilton, G., O'Connor, M., Nicholas, S., Marshall, R., Stanwell-Smith, R., Ndawula, E., Casemore, D., Gallagher, P. and Harnett, P. 1991. Cryptosporidiosis in the Isle of Thanet; an outbreak associated with local drinking water. *Epidemiological Infection*, 107 509.

Kang, P. K. and Shah, D. O. 1997. Filtration of nanoparticles with dimethyldioctadecylammonium bromide treated microporous polypropylene filters. *Langmuir*, 13 1820.

Kim, J., J., T. and Hall, M. 2003. Origins of filter particles: experimental study of particle deposition and detachment. *IWA International Conference*. Zurich, Switzerland:

Kim, J. K. and Tobiason, J. E. 2004a. Particles in filter effluent: the roles of deposition and detachment. *Environmental Science & Technology*,

Koo, T., Lee, Y. J. and Sheikholeslami, R. 2001. Silica fouling and cleaning of reverse osmosis membranes. *Desalination*, 139 43.

Landa, H., Capella, A. and Jimenez, B. 1997. Particle size distribution in an effluent from an advanced primary treatment and its removal during filtration. *Water Science and Technology*, 36 (4): 159.

Langlais, B., Reckhow, D. A. and Brink, D. R. 1991. Ozone in water treatment. AwwaRF and Compagnie Generale Des Eaux,

Lawler, D. F. 1997. Particle size distributions in treatment processes: theory and practice. *Water Science and Technology*, 36 (4): 15.

LeChevalier, M. and Norton, W. D. 1995. Giardia and Cryptosporidium in raw and finished water. *Journal American Water Works Association*, 54.

Leland, D., McAnulty, J., Keene, W. and Stevens, G. 1993. A Cryptosporidiosis outbreak in a filtered-water supply. *Journal American Water Works Association*, 85 (6): 34.

Lepore, J. V. and Ahlert, R. C. (1988) Fouling in membrane processes. In *Reverse Osmosis Technology*. Parekh, B. S. (ed). New York, NY: Marcel Dekker, p. 141.

Li, Q. G. and Elimelech, M. 2004. Organic fouling and chemical cleaning of nanofiltration membranes: measurements and mechanisms. *Environmental Science & Technology*, v38, p. 4683-4693.

Lisle, J. T. and Rose, J. B. 1995. Cryptosporidium contamination of water in the USA and UK: a mini-review. *Journal Water SRT-Aqua*, 44 (3): 103.

Logsdon, G. S., Symons, J. M., Hoye, R. L. and Arozarena, M. M. 1981. Alternate filtration methods for removal of Giardia cysts and cyst models. *Journal American Water Works Association*, 73 (2): 111.

Logsdon, G. S., Thurman, W. C., Frindt, E. S. and Stoecker, J. G. 1985. Evaluating sedimentation and various filter media for removal of Giardia cysts. *Journal American Water Works Association*, 77 (2): 61.

Mackenzie, W. R. 1994. A Massizve outbreak in Milwaukee of Cryptosporidium infection transmitted through the public service. *New England Journal of Medicine*, 331 (3): 161.

Mackie, R.I. and R. Bai, 1992. Suspended particle size distribution and the performance of deep bed filters. *Water Research*, **26**(12): p. 1571-1575.

Masciangioli, T. and Zhang, W.-X. 2003. Environmental Technologies at the nanoscale: Potentials & Pitfalls. *Environmental Science & Technology*, 37 (5), p102A-108A.

McDowell-Boyer, L. M. 1992. Chemical mobilization of micron-sized particles in saturated porous media under steady flow conditions. *Environmental Science & Technology*, 26 (3): 586.

McTigue, N. E., LeChavallier, M., Arora, H. and Clancy, J. 1998. National assessment of particle removal by filtration. International Waer Supply Association and AWWARF, Denver.

Moore, A. C., Herwaldt, B. L., Craun, G. F., Calderon, R. L., Highsmith, A. K. and Juranek, D. D. 1994. Waterborne disease in the United States, 1991 and 1992. *Journal American Water Works Association*, 86 (2): 87.

Muhammed, N., Ellis, K., Parr, J. and Smith, M. D. 1996. Optimization of slow sand filtration. *22nd WEDC Conference*. New Delhi, India:

Nieminski, E. C. and Ongerth, J. E. 1995. Removing Giardia and Cryptosporidium by conventional and direct filtration. *Journal American Water Works Association*, 96.

Nyström, M., Kaipia, L., Luque, S. 1995. Fouling and retention of nanofiltration membranes. *Journal of Membrane Science*, 98 (3): 249.

Nyström, M., Ruohomäki, K., Kaipia, L. 1996. Humic acid as a fouling agent in filtration. *Journal of Membrane Science*, 106 79.

O'Melia, C. R. and Stumm, W. 1967. Theory of Water Filtration. *J. A.W.W.A.*, 59 1393.

O'Melia, C. R. 1985. Particles, Pretreatment, and Performance in Water Filtration. *Journal of the Environmental Engineering, American Society of Civil Engineers*, 111 (6): 874.

O'Neill, M. E. 1964. A Slow Motion of Viscous Liquid Caused by a Slowly Moving Solid Sphere. *Mathematika*, 11 67.

Ongerth, J. E. and Pecoraro, J. P. 1995. Removing Cryptosporidium using multimedia filters. *Journal American Water Works Association*, 87 (12): 83.

Patania, N. L., Jacangelo, J. G., Cummings, L., Wilczak, A., Riley, K., Oppenheimer, J. and Watson, M. 1995. Optimization of filtration for cyst removal. AWWARF, Denver.

Patania, N. L., Jacangelo, J. G., Cummings, E. W., Wilczak, K., Riley, J., Oppenheim, I. and Watson, M. 1996. Optimization of filtration for cyst removal. AWWARF, Denver.

Raveendran, P. and Amirtharajah, A. 1995. Role of short-range forces in particle detachment during filter backwashing. *Journal of Environmental Engineering*, 860.

Rice, E. W., Fox, K. R., Miltner, R. J., Lytle, D. A. and Johnson, C. H. 1996. Evaluating plant performance with endospores. *Journal American Water Works Association*, 122.

Rimer, A. E. 1968. Filtration through a tri-media filter. *Journal of Environmental Engineering*, 94 (3): 521.

Roebeck, G. G., Dostal, K. A. and Woodward, R. L. 1964. Studies of modifications in water filtration. *Journal American Water Works Association*, 56 198.

Roefer, P., Monsevitz, J. and Rexing, D. J. 1995. The Las Vegas cryptosporidium experience. *AWWA Water Quality Technology Conference*. Denver, CO:

Roefer, P. A., Monscvitz, J. T. and Rexing, D. J. 1996. The Las Vegas Cryptosporidium outbreak. *Journal American Water Works Association*, 95.

Ryan, J. N. and Gschwend, P. M. 1994. Effects of Ionic Strength and Flow Rate on Colloid Release: Relating Kinetics to Intersurface Potential Energy. *Journal of Colloid and Interface Sci.*, 164 21.

Schäfer, A. I., Fane, A. G. and Waite, T. D. 1998. Nanofiltration of natural organic matter: removal, fouling and the influence of multivalent ions. *Desalination*, 118 109.

Smith, H. V., Patterson, W. J., Hardie, R., Greene, L. A., Benton, C., Tulloch, W., Gilmour, R. A., Girdwood, R. W. A., Sharp, J. C. M. and Forbes, G. I. 1989. An outbreak of waterborne Cryptosporidiosis caused by post-treatment contamination. *Epidemiological Infection*, 103 703.

Solo-Gabriele, H. and Neumeister, S. 1996. US Outbreaks of Crytosporidium. *Journal American Water Works Association*, 76.

Srinivasan, P. T., Viraraghavan, T. and Subramanian, K. S. 1999. Aluminum in drinking water: an overview. *Water SA*, 25 (1): 47.

Srinivasan, P. T. and Viraraghavan, T. 2002. Characterisation and concentration profile of aluminum during drinking-water treatment. *Water SA*, 28 (1): 99.

Swertfeger, J., Metz, D. H., DeMarco, J., Braghetta, A. and Jacangelo, J. G. 1999. Effect of filter media on cysts and oocyst removal. *Journal American Water Works Association*, 91 (9): 90.

Tchobanoglous, G. and Eliassen, R. 1970. Filtration of treated sewage effluent. *Journal of Sanitary Engineering*, 96 243.

Tien, C. 1989. *Granular Filtration of Aerosols and Hydrosols.* 1st Edition. Butterworths Series in Chemical Engineering, Boston.

Tobiason, J. E. and Vigneswaran, B. 1994. Evaluation of a modified model for deep bed filtration. *Water Research*, 28 (2): 335.

Tobiason, J. E. and Kim, J. 2001. Particles in Filter Effluent: Exploring the Roles of Removal and Detachment. *Advances in Granular Filtration in Water Treatment.* London, UK: International Conference, C. I. o. W. a. E. M.

Trussell, R., Trussell, A. R., Lang, J. S. and Tate, C. H. 1980. Recent developments in filtration system design. *Journal American Water Works Association*, 72 (12): 705.

Tufenkji, N., Miller, G. F., Ryan, J. N., Harvey, R. W. and Elimelech, M. 2004. Transport of Cryptosporidium Oocysts in porous media: Role of straining and physicochemical filtration. *Environmental Science & Technology*, 38(22): 5932.

Veerapaneni, S. and Wiesner, M. R. 1993. Role of Suspension Polydispersivity in Granular Media Filtration. *Jour. Envir. Engr., ASCE*, 119(1):172-190.

Veerapaneni, S. and Wiesner, M. R. 1998. Deposit Morphology and Head Loss Development in Porous Media *Environmental Science & Technology*, 31(10):2738-2744.

Wiesner, M. R. and Aptel, P. (1996) Mass transport, permeate flux, and fouling in pressure-driven processes. In *Membrane Processes for Water Treatment.* Mallevialle, J., Odendall, P. and Wiesner, M. R. (eds). New York: McGraw Hill.

Yiantsios, S. G. and Karabelas, A. J. 1998. The effect of colloid stability on membrane fouling. *Desalination*, 118 143.

Yohe, T. L., Heichel, J., Stromberg, B., Getting, T. M., Zukus, L. and Ball, C. 1999. The effect of low uniformity coefficient anthracite on dual-media filtration. *AWWA Annual Conference.* Denver, CO:

Yuksel, E., Akgiary, O., Saatci, A. M., Sarikaya, A. M. and Koyuncu, I. 2002. Effect of ozone injection location on filter performance in direct filtration. *Water Science and Technology*, 46 (9): 171.

Zhu, X. 1996. *Colloidal fouling of thin-film composite and cellulose acetate reverse osmosis membranes.* University of California, Los Angeles, Los Angeles.

Zhu, X. and Elimelech, M. 1995. Fouling of reverse osmosis membranes by aluminum oxide colloids. *Journal of Environmental Engineering*, 121 (12): 884.

Zhu, X. and Elimelech, M. 1997. Colloidal fouling of reverse osmosis membranes: measurements and fouling mechanisms. *Environmental Science and Technology*, 31 (12): 3654.

ABBREVIATIONS

AMSWTP	Alfred Merit Smith Water Treatment Facility
AWWA	America Water Works Association
AwwaRF	Awwa Research Foundation
ºC	Degress celsius
CV	coefficient of variation
d	diameter
EDX	Energy dispersive X-Ray spectrometer
EDAX	Energy dispersive analysis by X-ray
EDS	Energy Dispersive Spectrometer
USEPA	United States Environmental Protection Agency
EWPP	East water purification plant, Houston, TX
ft	foot
GAC	Granular activated carbon
gpm	gallons per unit
HCl	Hydrochloric acid
in	inch
ICP	Inductively Coupled Plasma
mgd	million gallons per day
mg/L	milligram/liter
min	minute
MRWTP	Modesto Regional Water Treatment Plant
MS	Mass spectrometer
μg	micrograms
μm	micrometer (=10^{-6} m)
nm	nanometer (=10^{-9} m)
NTU	Napthlometric turbidity units
OES	Optical Emission Spectrometer
PAC	Powdered activated carbon
PACl	Polyaluminum chloride
ppb	parts per billion (in this study refers to μg/L)
ppm	parts per million (in this study refers to mg/L)
rpm	revolutions per minute

RMWTP	Richard Miller Water Treatment Plant, Greater Cincinnati Water Works
SEM	Scanning Electron Microscopy
TEM	Transmission Electron Microscopy
TOC	Total organic carbon
UC	Uniformity coefficient

IWA Publishing's authorised EU representative for General Product Safety Regulations is Diane D'Arras, 15 rue Duret, 75116 Paris, France, e-mail: safety@iwap.co.uk.

Printed and bound by CPI Group (UK) Ltd, Croydon, CR0 4YY
17/09/2026
02218808-0002